Jacek Mańdziuk

Knowledge-Free and Learning-Based Methods in Intelligent Game Playing

Studies in Computational Intelligence, Volume 276

Editor-in-Chief

Prof. Janusz Kacprzyk
Systems Research Institute
Polish Academy of Sciences
ul. Newelska 6
01-447 Warsaw
Poland
E-mail: kacprzyk@ibspan.waw.pl

Further volumes of this series can be found on our
homepage: springer.com

Vol. 257. Oscar Castillo, Witold Pedrycz, and
Janusz Kacprzyk (Eds.)
*Evolutionary Design of Intelligent
Systems in Modeling, Simulation and Control,* 2009
ISBN 978-3-642-04513-4

Vol. 258. Leonardo Franco, David A. Elizondo, and
José M. Jerez (Eds.)
Constructive Neural Networks, 2009
ISBN 978-3-642-04511-0

Vol. 259. Kasthurirangan Gopalakrishnan, Halil Ceylan, and
Nii O. Attoh-Okine (Eds.)
*Intelligent and Soft Computing in Infrastructure Systems
Engineering,* 2009
ISBN 978-3-642-04585-1

Vol. 260. Edward Szczerbicki and Ngoc Thanh Nguyen (Eds.)
Smart Information and Knowledge Management, 2009
ISBN 978-3-642-04583-7

Vol. 261. Nadia Nedjah, Leandro dos Santos Coelho, and
Luiza de Macedo de Mourelle (Eds.)
Multi-Objective Swarm Intelligent Systems, 2009
ISBN 978-3-642-05164-7

Vol. 262. Jacek Koronacki, Zbigniew W. Ras,
Slawomir T. Wierzchon, and Janusz Kacprzyk (Eds.)
Advances in Machine Learning I, 2009
ISBN 978-3-642-05176-0

Vol. 263. Jacek Koronacki, Zbigniew W. Ras,
Slawomir T. Wierzchon, and Janusz Kacprzyk (Eds.)
Advances in Machine Learning II, 2009
ISBN 978-3-642-05178-4

Vol. 264. Olivier Sigaud and Jan Peters (Eds.)
*From Motor Learning to Interaction
Learning in Robots,* 2009
ISBN 978-3-642-05180-7

Vol. 265. Zbigniew W. Ras and Li-Shiang Tsay (Eds.)
Advances in Intelligent Information Systems, 2009
ISBN 978-3-642-05182-1

Vol. 266. Akitoshi Hanazawa, Tsutom Miki,
and Keiichi Horio (Eds.)
Brain-Inspired Information Technology, 2009
ISBN 978-3-642-04024-5

Vol. 267. Ivan Zelinka, Sergej Celikovský, Hendrik Richter,
and Guanrong Chen (Eds.)
Evolutionary Algorithms and Chaotic Systems, 2009
ISBN 978-3-642-10706-1

Vol. 268. Johann M.Ph. Schumann and Yan Liu (Eds.)
Applications of Neural Networks in High Assurance Systems,
2009
ISBN 978-3-642-10689-7

Vol. 269. Francisco Fernández de de Vega and
Erick Cantú-Paz (Eds.)
Parallel and Distributed Computational Intelligence, 2009
ISBN 978-3-642-10674-3

Vol. 270. Zong Woo Geem
Recent Advances In Harmony Search Algorithm, 2009
ISBN 978-3-642-04316-1

Vol. 271. Janusz Kacprzyk, Frederick E. Petry, and Adnan
Yazici (Eds.)
*Uncertainty Approaches for Spatial Data Modeling and
Processing,* 2009
ISBN 978-3-642-10662-0

Vol. 272. Carlos A. Coello Coello, Clarisse Dhaenens, and
Laetitia Jourdan (Eds.)
Advances in Multi-Objective Nature Inspired Computing,
2009
ISBN 978-3-642-11217-1

Vol. 273. Fatos Xhafa, Santi Caballé, Ajith Abraham,
Thanasis Daradoumis, and Angel Alejandro Juan Perez
(Eds.)
*Computational Intelligence for Technology Enhanced
Learning,* 2010
ISBN 978-3-642-11223-2

Vol. 274. Zbigniew W. Raś and Alicja Wieczorkowska (Eds.)
Advances in Music Information Retrieval, 2010
ISBN 978-3-642-11673-5

Vol. 275. Dilip Kumar Pratihar and Lakhmi C. Jain (Eds.)
Intelligent Autonomous Systems, 2010
ISBN 978-3-642-11675-9

Vol. 276. Jacek Mańdziuk
*Knowledge-Free and Learning-Based Methods in Intelligent
Game Playing,* 2010
ISBN 978-3-642-11677-3

Jacek Mańdziuk

Knowledge-Free and Learning-Based Methods in Intelligent Game Playing

 Springer

Prof. Jacek Mańdziuk
Faculty of Mathematics and
Information Science
Warsaw University of Technology
Plac Politechniki 1
00-661 Warsaw
Poland
E-mail: J.Mandziuk@mini.pw.edu.pl

ISBN 978-3-642-26213-5   e-ISBN 978-3-642-11678-0

DOI 10.1007/978-3-642-11678-0

Studies in Computational Intelligence   ISSN 1860-949X

Typeset & Cover Design: Scientific Publishing Services Pvt. Ltd., Chennai, India.

Printed in acid-free paper

9 8 7 6 5 4 3 2 1

springer.com

To my wife Magda,
and children: Zosia, Joasia and Tomek

Foreword by David B. Fogel

At a time when biologically inspired machine intelligence tools are becoming crucial to the development of game-playing strategies, Jacek Mańdziuk has written a book that focuses on exactly these tools. His book *Knowledge-free and Learning-based Methods in Intelligent Game Playing* concentrates on games requiring mental ability "mind games" as Mańdziuk describes them. His extensive survey of the literature is aimed mainly on those methods that offer the potential to have a computer teach itself how to play a game starting with very little (or no) knowledge of the game, or to learn how to improve its play based on feedback about the quality of its performance. His perspective is one of a futurist, looking ahead to grand challenges and the computational intelligence methods that may be used to meet those challenges.

To date, the greatest achievements in game play in most standard "mind games" have come from knowledge-intensive methods, not knowledge-free methods. They have come from programming human knowledge into the software, rather than having the software learn for itself. This is true, for example, for Othello, checkers, and chess.

But this is going to have to change because the games that now really challenge us are sufficiently complicated that traditional knowledge-intensive methods will not yield satisfactory results. Computing power alone is unlikely to be the basis for addressing Go, and even if that computational power does become sufficient, we could easily imagine Go on a 190×190 board instead of the 19×19 traditional game, which our modern computers already find vexing. In that case, Monte Carlo tree searches to the end of the game, endgame databases, and opening books based on human grand master knowledge would no longer be very helpful to the computers.

When compared to Go, the games than we humans play in the real world are of far greater complexity. For example, we could look to real-time strategy video games that have thousands of "agents" acting simultaneously, but we could just as well consider, say, the games of economics that are played between nations. Determining which countries should receive favorable trading status and which should have tariffs placed on their goods is a game. Its not played on a board or a table or a computer, but it is game nonetheless, and the consequences of winning or losing are usually far greater than is found in a contest between, say, two checkers players. I wonder if anyone truly believes that the usual devices of traditional artificial intelligence will ever be very successful in addressing such games. Certainly, I do not believe they will be.

An alternative is to focus on how humans adapt to different challenges, or more generally how any intelligent systems do this. Intelligence then becomes a matter of adapting behavior to meet goals across a range of environments, and we can observe examples not only ourselves, but also in societies (such as ants) and in evolving organisms. Neural networks, evolutionary algorithms, and reinforcement learning take center stage, and here Mańdziuk highlights their application to games including backgammon, checkers, and Othello, with additional remarks toward poker, bridge, chess, Go, and other games of strategy. His perspective is that the key to making major advancements in intelligent game playing comes from capturing our human abilities such as creativity, selective focus, and generalization. His conviction is compelling.

This is a book that is both broad and detailed, comprehensive and specific, and it offers a vision toward the challenges that await us. With the expanding interest in game playing Mańdziuk offers a timely book that will serve the growing games community for years to come. I hope you find it a source of continual inspiration.

David B. Fogel
Lincoln Vale CA LP
Natural Selection, Inc.
San Diego, California, USA

Foreword by Simon M. Lucas

Since the seminal work of the founding fathers of computer science and information theory, people such as Alan Turing and Claude Shannon, there has been a rich history of research applied to games, and without doubt this has contributed greatly to the development of artificial intelligence as a discipline. The focus of this book is on Mind Games, games of skill such as Chess, Go, and Bridge which predate computers. There have been two approaches to designing agents capable of a high standard of play in such games: hand programming, and machine learning. Both methods have had their high-profile successes, with Deep Blue (Chess) being an exemplar of expert hand-tuning, and TD-Gammon (Backgammon), Logistello (Othello) and Blondie (Checkers) being exponents of the machine-learning approaches (though each of these differed in the nature of the learning). As the book argues, there is much more to this field than simply designing agents that are capable of a high standard of play. The way in which it is achieved is of great importance. Hand programmed agents tend to be specific to a particular game. Learning agents on the other hand offer the potential of a more general ability, and the ability to learn is central to most meaningful definitions of intelligence. Games offer an excellent application domain and a demanding test-bed for AI and CI techniques.

This book offers a unique perspective on the subject, embracing both conventional AI approaches to game-playing as well as the open-ended challenges that games pose for computational intelligence. It gives a detailed account of the main events in the development of the field, and includes the most important algorithms in game-tree search, temporal difference learning, and in evolutionary game learning, and describes in detail how these have been applied to a range of different mind games. The focus on mind games enables a deep coverage of the most important research in this area.

The book is organised into four parts. The first part covers the foundations including the main game-tree search algorithms and then discusses the state of the art in the most important mind games. Part two describes the main computational intelligence methods that have been applied to learn game strategy, and describes selected TD and neuro-evolutionary approaches to specific games including Backgammon, checkers, and Othello. Part three then goes into greater detail on how to get the best out of these methods, including the choice of function approximation architecture, the selection of game-features, and tuning the algorithms for best performance. It also discusses the main approaches to move ordering, and to opponent modelling. Part four describes some of the grand challenges in game playing, organised into chapters on intuition, creativity and knowledge discovery, and general game playing.

This is a well balanced book that celebrates the great achievements of the field while pointing out that many basic questions have yet to be answered. At all times the author shows an impressive knowledge of the subject and an ability to communicate this clearly. The book makes fascinating reading and should be of great interest to researchers involved in all kinds of computational intelligence and AI, and will help attract more people to study games. For PhD students starting out in the area it should be regarded as essential, and would also be useful as recommended reading for taught courses in computational intelligence and games.

Professor Simon M. Lucas
University of Essex, UK
Editor-in-Chief
IEEE Transactions on Computational Intelligence and AI in Games

Preface

Humans and machines are very different in their approaches to game playing. Humans use intuition, perception mechanisms, selective search, creativity, abstraction, heuristic abilities and other cognitive skills to compensate their (comparably) slow information processing speed, relatively low memory capacity, and limited search abilities. Machines, on the other hand, are extremely fast and infallible in calculations, capable of effective brute-force-type search, use "unlimited" memory resources, but at the same time are poor at using reasoning-based approaches and abstraction-based methods. The above major discrepancies in the human and machine problem solving methods underlined the development of traditional machine game playing as being focused mainly on engineering advances rather than cognitive or psychological developments. In other words, as described by Winkler and Fürnkranz [347, 348] with respect to chess, human and machine axes of game playing development are perpendicular, but the most interesting, most promising, and probably also most difficult research area lies *on the junction between human-compatible knowledge and machine compatible processing.* I undoubtedly share this point of view and strongly believe that the future of machine game playing lies in implementation of human-type abilities (abstraction, intuition, creativity, selective attention, and other) while still taking advantage of intrinsic machine skills.

The book is focused on the developments and prospective challenging problems in the area of mind game playing (i.e. playing *games that require mental skills*) using Computational Intelligence (CI) methods, mainly neural networks, genetic/evolutionary programming and reinforcement learning.

The majority of discussed game playing ideas were selected based on their *functional similarity* to human game playing. These similarities include: learning from scratch, autonomous experience-based improvement and example-based learning. The above features determine the major distinction between CI and traditional AI methods relying mostly on using effective game tree search algorithms, carefully tuned hand-crafted evaluation functions or hardware-based brute-force methods.

On the other hand, it should be noted that the aim of this book is by no means to underestimate the achievements of traditional AI methods in game playing domain. On the contrary, the accomplishments of AI approaches are undisputable and speak for themselves. The goal is rather to express my belief that other alternative ways of developing mind game playing machines are possible and urgently needed.

Due to rapid development of the AI/CI applications in mind game playing, discussing or even mentioning all relevant publications in this area is certainly beyond the scope of the book. Hence, the choice of references and presented methods had to be restricted to a representative subset, selected subjectively based on my professional knowledge and experience.

The book starts with the Introduction followed by 14 chapters grouped into 4 parts. The introductory chapter provides a brief overview of the book's subject, further elaborating on the reasons behind its origin and discussing its content in more detail. The first part of the book entitled *AI tools and state-of-the-art accomplishments in mind games* covers a brief overview of the field, discusses basic AI methods and concepts used in mind game research and presents some seminal examples of AI game playing systems.

Part II - *CI methods in mind games. Towards human-like playing* starts with an overview of major subdisciplines within CI, namely neural networks, evolutionary methods, and reinforcement learning, in the context of game playing developments. The other chapter in Part II emphasizes the selected accomplishments of Computational Intelligence in the field.

The third part - *An overview of challenges and open problems* presents, chapter-by-chapter, five aspects of mind game playing, which I consider challenging for CI methods. These areas are related to evaluation function learning, efficient game representations, effective training schemes, search-free playing and modeling the opponent. The topics are presented in the context of achievements made to date, open problems, and possible future directions.

The final part of the book, entitled *Grand challenges*, is devoted to the most challenging problems in the field, namely: implementation of intuitive playing, creativity and multi-game playing. All three above-mentioned issues are typical for human approaches to mind games and at the same time extremely hard to implement in game playing machines - partly due to not yet fully uncovered mechanisms of perception, representation and processing of game information in the human brain. The last chapter highlights the main conclusions of the book and summarizes prospective challenges.

In summary, I believe that the need for further development of human-like, knowledge-free methods in mind games is unquestionable, and the ultimate goal that can be put forward is building a truly autonomous, human-like multi-game playing agent. In order to achieve this goal several challenging problems have to be addressed and solved on the way. Presentation of these challenging issues is the essence of this book. I hope the book will be helpful, especially for young scientists entering the field, by summarizing the state-of-the-art accomplishments and rising new research perspectives.

I would like to thank Prof. Janusz Kacprzyk for the inspiration for writing this book and Prof. Włodzisław Duch for fruitful discussions concerning the scope of the book and related subjects. I'm very grateful to Prof. David Fogel and Prof. Simon Lucas for writing forewords and for encouraging comments regarding the book's scope and content. I would also like to thank my former and current PhD students: Cezary Dendek, Krzysztof Mossakowski, Daniel Osman and Karol Walędzik, as well as other co-authors of my papers discussed in this book for a satisfying and productive cooperation. I wish to thank Karol Walędzik for his valuable comments and suggestions concerning the initial version of the book. Last but not least I'm grateful to my family for supporting me in this undertaking.

Warsaw,
September 2009   Jacek Mańdziuk

Contents

1

Introduction

Playing games has always been an important part of human activities and the oldest mind games[1] still played in their original form (Go and backgammon) date back to 1,000 - 2,000 BC.

Games also became a fascinating topic for Artificial Intelligence (AI). The first widely-known "AI approach" to mind games was noted as early as 1769 when Baron Wolfgang von Kempelen's automaton chess player named *The Turk* was presented at the court of Empress Maria Theresa. *The Turk* appeared to be a very clever, actually unbeatable, chess player who defeated Napoleon and the Empress Catherine of All the Russias among others. It took a few decades to uncover a very smart deception: a world-class human player was hidden inside the Turk's machinery and through a complicated construction of levers and straddle-mounted gears was able to perceive opponent's moves and make its own ones. This fascinating history was described by several authors, including Edgar Allan Poe [256], the famous American novelist. Although *The Turk* had apparently nothing in common with AI, the automaton is an evident illustration of humans' perennial aspiration for creating intelligent machines able to defeat the strongest human players in popular mind games.

The first genuine attempt at artificial chess playing was made around 1890 by Torres y Quevedo who constructed a device capable of playing king and rook versus king endings (the machine was playing the more powerful side). The machine check-mated the opponent regardless of his/her play. Since the problem is relatively simple - a rule-based algorithm can easily be developed for endings of this type - the device may possibly be not regarded as sophisticated achievement according to today's standards, but was definitely advanced for the period of its origin.

Serious, scientific attempts to invent "thinking machines" able to play mind games began in the middle of the previous century. Thanks to seminal papers devoted to programming chess [238, 296, 327] and checkers [277] in the

---

[1] Throughout the book the term *mind games* will be used as a synonym of *games requiring mental skills*.

J. Mańdziuk: Knowledge-Free and Learning-Based Methods, SCI 276, pp. 1–7.
springerlink.com   © Springer-Verlag Berlin Heidelberg 2010

1950s, games remained through decades an interesting topic for both classical AI and CI based approaches.

One of the main reasons for games' popularity in AI/CI community is the possibility to obtain cheap, reproducible environments suitable for testing new search algorithms, evaluation methods, or learning concepts. Games can be tailored towards specific research questions, like behavior in one-, two- or multi-agent environment, analysis of perfect- vs. imperfect-information decision methods, optimal policy learning in zero-sum games vs. arbitrary-payoff ones, etc. From the engineering and scientific point of view games provide a flexible test bed for posing and answering questions within the above frameworks.

On the other hand, the "social factor" of game playing should not be underestimated, either. People always tended to challenge themselves in their mental skills, and one of the ways to make such a contest interesting in the domain of games was creating high-profile artificial mind game players.

Certainly, human interests in computer game playing extend far beyond the mind game area. Many other types of games, e.g. real-time strategy (RTS), skill, adventure, strategic, war, sport, negotiating, racing, and others are currently of rapidly growing interest, as well. Each of the above-mentioned types of games has many enthusiastic players and deserves separate, individual recognition. Each type of computer games has its own specificity, characteristic features and challenging problems. In the RTS games, for example, there are hundreds of moving pieces to be handled, each of them with a frequency of a few times per second. Each piece has generally only local information about its neighborhood, restricted to its sensing capabilities. Such a task is entirely different from the case of controlling (playing) perfect-information, zero-sum board games, in which the environment is discrete, deterministic, and updated synchronously in a step-by-step manner.

Generally speaking, the focus of this book is on the classical and most popular mind board games, i.e. chess, checkers, Go, and Othello[2]. The reason for choosing these particular games is twofold. First of all, these games are very popular and played all over the world. Secondly, for decades they have been a target for AI/CI research aiming at surpassing human supremacy. Even though the machines have already outperformed the top human players in all of them but Go, they are still interesting test beds for making comparative tests of new ideas, including the CI-related approaches. A very different specificity of each of these games allows for testing the whole spectrum of methods, from very specialized, usually game-related AI approaches, through universal search techniques, to completely general, game-independent issues related to universal learning schemes, pattern recognition, or knowledge discovery. For the sake of completeness of the book some references to another widely popular board game - backgammon and the two most popular card games - poker and bridge are also included in the book.

---

[2] As a rule, all names of the games are written in the book with non-capital letters. The only exceptions are the games which are written commonly with the initial capital, like Go, Othello, Scrabble, or Perudo.

Certainly, without any doubt there are many other interesting and highly competitive mind games (e.g. shogi, Chinese chess, hex, Amazons, Octi, lines of actions, sokoban), which would be worth considering in this book. The main reason for not taking them into account is their still relatively lesser popularity - although, some of them are becoming more and more prominent. There are, however, two exceptions from this popularity-based selection. Due to the author's personal interests, remarks concerning two less popular games: give-away checkers and Perudo (Liar's Dice) are also included in the book.

Since the four main target games (chess, checkers, Othello, and Go) are very popular all over the world, it is assumed in the book that the rules of these games are known to the reader. In case this assumption is incorrect the appropriate references are given in the following paragraphs. The four target games can be compared briefly based on a few key parameters: the number of possible states, the average branching factor, and the average length of the game. Certainly, these numbers alone do not provide complete characteristics of these games, but they can be used as first, crude estimation of their complexity.

Chess - From the very beginning of AI involvement in games, it was chess - the queen of mind games at that time[3] - that attracted special attention. In 1965 the game was even announced "the Drosophila of Artificial Intelligence" by the Russian mathematician Alexander Kronrod. The rules of chess can be found in various publications, including official web pages of national chess associations, e.g. United States Chess Federation [328]. The branching factor in chess equals approximately 35, the average length of a chess game is about 40 moves (80 half-moves). Hence the game tree complexity is around 10^{123} (35^{80}). The number of possible board configurations equals about 10^{50} [296].

Checkers - is the first widely-known and sufficiently complicated game that was solved, i.e. the result of the game is known assuming perfect play of both sides. In such a case the game leads to a draw. The rules of checkers can be found for example on the American Checkers Federation WWW site [6]. The average branching factor equals 3 - a relatively low value is attributed to the fact that all jumps are forced. The state space complexity exceeds 5×10^{20} [284]. The average length of a game is around 35 moves (70 ply). Hence the estimated game tree complexity is around 10^{31}.

Othello - is simpler than chess, but more demanding than checkers. The average branching factor of Othello equals 7 [184]. The average length of a game exceeds 30 moves (60 ply) since the game is played until the board is complete or none of the sides can make a legal move. The number of legal positions in Othello is approximated by 10^{28} [5] and the size of a game-tree by 10^{50}. Even though the rules of making moves in Othello are very simple [45] the game itself is definitely far from trivial. The two most significant distinctions between Othello and chess or checkers is its additive nature (in Othello, at each turn a new piece is added to the board and all played pieces remain until the end of the game) and high volatility of the temporal score

---

[3] Nowadays many will probably argue for assigning this title to Go.

(since at each turn a relatively high number of pieces may be flipped, the actual score may - and often does - change dramatically in a single turn).

Go - is undoubtedly the most challenging game for AI/CI methods among the four considered and one of the most demanding games in general. Even in a simplified 9×9 version the machine players are still below top humans, not mentioning the 19×19 version of the game. The branching factor of Go is estimated at a value of a few hundred. Assuming it to be equal to 250, with the average game length of about 150 ply, the game tree complexity is roughly equal to 10^{360}, being enormous number for any computing machines possibly imagined. The state space complexity of Go is bounded by 3^{361} ($\approx 10^{172}$), which is again beyond the reach of any exhaustive search method. Introduction to the game of Go can be found in [69] or on several Internet sites. For discussion on computational aspects of the game please refer to [237]. Except for a very high branching factor, the specificity of Go lies in difficulty of efficient evaluation of the board position. Such an evaluation must include several subtle trade-offs and analysis of possible variants far into the future. The overall value of Go position is a combination of local assessments of individual groups of stones, which need to be classified as being alive or not, as having or not having chances for making strong connections with other stones, etc.

Even with a quite restricted choice of games used to illustrate the main ideas presented in this book, there is no practical possibility to discuss or even mention all relevant related papers and discuss all interesting issues. With regard to the book's organization the basic decision was whether to use paper-by-paper presentation, focusing on some number of the most significant and representative publications, or organize the book around a limited number of the most relevant CI-related game issues. The decision was to use the latter organizational layout, which has an advantage over the former in making the presentation of each issue coherent and as in-depth as necessary. The consequence of this choice, however, is that some of the relevant papers, which are not directly related to any of the issues discussed in the book are cited in a general context only. On the other hand, due to the above-mentioned convention, the content of the paper which fits to more than one topic is presented and discussed (in part) in several, respective chapters.

Before summarizing the book's content a distinction between AI and CI methods needs to be clarified. Naturally, there is no widely accepted definition of AI versus CI and the author has no ambition to exactly draw this thin, partly subjective line between these two fields. Instead, it is proposed to use any of the following three "operational" definitions.

First, the tools and techniques used by classical AI and CI are to large extent different: references to CI systems usually address soft-computing-based methods *which are inspired from nature*, i.e. neural networks [33, 147], evolutionary methods [110, 140, 150], fuzzy systems [239, 350, 352], reinforcement learning [163, 313], swarm intelligence [83, 87], or rough sets [245, 246].

Second, unlike traditional AI solutions, the CI-based systems are *capable of learning and autonomous improvement of behavior*. The learning paradigm can be implemented either as a guided or an unguided process, e.g. as example-based neural network training or self-playing reinforcement-based training. The CI learning also includes knowledge-free approaches and learning from scratch in a lifelong horizon.

The third possible distinction is related to the nature of intelligence in AI vs. CI systems: "In AI research the emphasis is on producing apparently intelligent behavior using whatever techniques are appropriate for a given problem. In CI research, the emphasis is placed on intelligence being an emergent property." [195].

Regardless of the nuances concerning definitions of AI and CI areas, the fact is that currently the AI/CI-related research in mind games is developing rapidly. In 2006 the author was writing a book chapter discussing challenges to Computational Intelligence in mind games [206]. As a part of this work the state-of-the-art AI/CI achievements in mind games were listed. Since then several new world-class developments in this area have emerged, e.g. the chess super program Rybka [105, 261] or the MoGo [126, 340] program in Go. Also high advancement of Quackle [166] - a Scrabble playing program or ChessBrain Project [116] was observed, to name only a few achievements.

Recent advances of AI in the most popular mind games, which led to spectacular challenges to the human supremacy in chess, checkers, Othello, or backgammon provoke the question: *Quo vadis mind game research?*. What may be the reason for pursuing the CI research in this area? For example, what else can possibly be gained in checkers, which is already a solved game? Can CI-based methods raise the bar in chess higher than had already been done by Deep Blue and further advanced by Rybka? Do we still need to pursue mind game research or maybe defeating human world champions in all major board games but Go is (was) the ultimate, satisfying goal? Well, the answers to the above principal questions are certainly not straightforward. Obviously, when considering the quality of machine playing in a particular game as the sole reference point the only remaining target might be further extension of the machines' leading margin in the man-machine competition (since it is doubtful that the improvement of human players will be adequate to that of computer players).

But improvement of efficiency is not the only and not even a sufficient motivation for further research. In the author's opinion good reasons for game research concern *the way* in which high playing competency is accomplished by machines [207]. On one side, there are extremely powerful AI approaches in which playing agents are equipped with carefully designed evaluation functions, look-up tables, perfect endgame databases, opening databases, grandmaster game repositories, sophisticated search methods and a lot of other predefined, knowledge-based tools and techniques that allow making high quality moves with enormous search speed. On the other side, there are soft, CI-based methods relying mainly on knowledge-free approaches, extensive training methods including reinforcement learning and neural networks, self-playing, and even learning from scratch based merely on the final outcomes of the games played. In the author's opinion there is room for further development of these human-type methods of perception, learning, and reasoning. The search for a truly intelligent, autonomous *thinking machine* can be in part realized within the game community - mostly with the help of Computational Intelligence. Further development and successful application of the above techniques pose several challenging questions which are the focus of this book.

The remaining content of the book is divided into four parts. Part I composed of three chapters, entitled *AI tools and state-of-the-art accomplishments in mind games*, is dedicated to traditional AI approaches in games (AIG). It starts (chapter 2) with a brief historical note about the beginning of the AIG discipline with particular emphasis on the ideas presented by Claude Shannon in his highly influential article [296]. Chapter 3 presents selected methods and techniques used in AIG, in particular game representation in the form of a game tree, the minimax search method, alpha-beta pruning algorithm with various modifications (history heuristics, transposition tables, iterative deepening, killer moves, etc.), SCOUT, Negascout, MTD(f) and MTD(f)-bi search algorithms, and Monte Carlo sampling-based simulations. The last chapter of the first part presents the greatest achievements of AIG in the most popular games. First, the history of AI in chess is presented, starting as early as the 18th century's baron von Kempelen's Turk automaton, up to the state-of-the art Rybka program. Next Chinook - the perfect checkers playing program written by Jonathan Schaeffer and his collaborators - is introduced, followed by presentation of Logistello - the former Othello World Champion program written by Michael Buro, and the two well-known Scrabble programs: Maven and Quackle - both attaining the world class level of play. The remaining three games considered in chapter 4 are bridge, poker and Go - all of them being out of artificial player's reach yet.

Part II entitled *CI methods in mind games. Towards human-like playing* is composed of two chapters. The first one briefly describes the most popular CI methods used in mind game domain, namely neural networks, evolutionary methods, and reinforcement learning. The basic concepts underlying each of these fields are first introduced, followed by an application overview in the game domain context. For the sake of clarity of the presentation a more detailed description of particular applications is left for the following chapters, each of which is devoted to a specific challenging issue. In the second chapter of Part II selected CI-based approaches to mind game playing are presented. Some of them, like Gerald Tesauro's TD-Gammon or David Fogel's Blondie24 are seminal achievements in the field, some other may be less acknowledged, but each of them provides some new, interesting ideas. The descriptions of these accomplishments are concise, focused mainly on highlighting representational, methodological and computational aspects of proposed approaches.

Chapters 7 - 11 constitute the third part of the book entitled *An overview of challenges and open problems*. Each chapter is dedicated to one particular issue: in chapter 7 the problem of autonomous evaluation function learning is discussed. In its simpler form the problem boils down to learning the appropriate weight coefficients in the linear combination of predefined game features. More challenging formulation of the evaluation function learning problem relies on autonomous specification of the set of atomic features that compose this linear weighted combination. In the case of a nonlinear evaluation function the natural candidate is neural network representation. The next chapter presents the significance of game representation in CI related systems respectively for checkers, Go, Othello, and bridge. Chapter 9 highlights the importance of the training schemes used in CI systems, with emphasis on temporal difference learning. In particular, the self-playing mode is compared with playing against external opponents. In the latter case the problem of optimal selection of the opponents is considered. In this context several training strategies, e.g. learning exclusively on games lost or drawn vs. learning on all games, or choosing each of the opponents with equal frequency vs. playing more often with stronger opponents, etc. are compared. A discussion on the efficacy of temporal difference learning vs. evolutionary methods concludes the chapter. Chapter 10 touches a very practical issue of efficient moves pre-ordering in the context of search-free or shallow search-based playing systems. The last chapter in the third part of the book covers the problems related to modeling the opponent's playing style and modeling uncertainty in games. Although these issues have only limited use in perfect-information games, there is still some possibility to explore this subject in the context of modeling the strengths and weaknesses of a particular playing style (e.g. tactical vs. positional playing in chess). The most straightforward applications of opponent modeling are related to games allowing deception or bluffing, in which modeling the current opponent's "mood" as well as his general, long-term playing pattern (e.g. aggressive vs. conservative) is considered a challenging problem. The problem of modeling the opponent in bluffing games is illustrated on the example of Perudo (also known as Liar's Dice) and poker.

The last part of the book - entitled *Grand challenges* - is a quintessence of the human-like playing advocated in this monograph. Three major goals on the path to human-type playing intelligence are presented in the three subsequent chapters. These fundamental, challenging issues are: intuitive playing, creativity, and multi-game playing. All these three abilities are easily, almost effortlessly accomplishable for human players and, at the same time, still far from being captured by machines. Among these three wonderful human abilities, it is intuition that is the most mysterious and the most difficult to implement in machine playing, although some attempts to harness this specifically human aptitude in game playing context have been reported in the literature.

The last chapter of the book summarizes its main theses and sketches possible scenarios for future development of the Computational Intelligence methods in mind game playing.

Part I

AI Tools and State-of-the-Art
Accomplishments in Mind Games

Part I

Foundations of AI and CI in Games. Claude Shannon's Postulates

From the very beginning of Artificial Intelligence, games were one of the main topics of interest and extensive research. The foundations of the field were laid in seminal papers by the giants of AI and general computer science: von Neumann and Morgenstern [333], Shannon [296], Turing [327], Newell, Shaw and Simon [238], Samuel [277, 279], and others. In the 60 years that followed these first investigations, all relevant aspects of machine game playing were developed step by step, ultimately leading to world championship level playing systems. Apart from these top AI/CI achievements there are plenty of other less renown, but relevant research results. Several search methods, smart game representations, sophisticated evaluation functions, and other game related concepts were developed on the way. A selection of these accomplishments is presented in the following two chapters devoted respectively to AI methods and AI playing systems. The reader interested in systematic and exhaustive presentation of AI developments in particular games may be willing to consult overview papers, e.g. [120, 303] for chess or [40, 237] for Go.

Roughly speaking the development of AI/CI in games may be divided into four phases: the early fascination of the field in the 1940s and 1950s, the breakthrough caused by the invention of alpha-beta search method and related discoveries in the 1960s and 1970s, the period between mid-1970s and mid-1990s, in which brute force search took the lead, and the last $10-15$ years, in which several fundamental questions about the future of AI/CI in games have been put forward and the community has tried to redefine the goals after Deep Blue's victory over Kasparov and - very recently - solving the game of checkers.

This book tries to join this discussion and sketch the possible future directions and challenging problems in which - in the author's opinion - quite a few relevant open questions still remain. Such a discussion on human-like, trainable playing systems *must* start with a short reminder of Shannon's paper [296].

Shannon proposed a constructive way of building the evaluation function, whose underlying aspects have remained applicable in the game playing

J. Mańdziuk: Knowledge-Free and Learning-Based Methods, SCI 276, pp. 11–13.
springerlink.com   © Springer-Verlag Berlin Heidelberg 2010

programs through decades. The function is constructed as a weighted sum of material and particular positional features and each element of the sum is defined as the difference between the player's and his opponent's values. Certainly the key issue is how to define the weights of such a linear combination and how to define the relevant features to be represented in the function. The latter question is especially important with respect to positional features, which are more elusive and harder to define in the form of closed and representative set than components representing material advantage.

The other main topic covered by [296] is the problem of efficient move selection. Shannon proposed two playing strategies denoted by A and B, respectively. In the context of game tree searching the former one relied on searching the whole game tree to some predefined depth and was essentially an implementation of the minimax algorithm (discussed in detail in section 3.2.1). Shannon concluded that a computer implementing strategy A would be both a slow and weak player since the time restrictions would only allow for a shallow, few-ply-deep search. Shannon estimated that under very optimistic assumptions about the computer processing speed the achievable search depth would be equal to 3 half-moves. Moreover, a uniform search with fixed depth limit might possibly end-up in the middle of material exchange and the evaluation value returned by the program in such a non-quiescent position would be pointless.

Considering the above, Shannon proposed an extension of the above search procedure - called strategy B - which operates based on two general principles:

- examine forceful moves as deeply as possible (given time constraints) and execute the evaluation procedure only at quiescent or quasi-quiescent positions,
- explore only selected variations and assign priorities to the moves: the highest values are assigned to forceful and semi-forceful as well as capture and threatening moves (e.g. giving check to the opponent's king or attacking an opponent's high-valued piece by our lower valued one), the medium relevance is given to defensive moves, and low priorities to other moves.

The algorithms of type B are definitely "smarter" than those implementing strategy A since they explore only certain subset of all possible variations with special care devoted to the promising continuations. On the other hand, in the above formulation the division between promising and non-promising moves is defined based on quite general assumption of the advantage of forceful and attacking moves over the defensive ones and over the "neutral" moves. Although this assumption is generally true, in several game situations more selective investigation is desirable. Hence, Shannon proposed a variation of the strategy B which relies less on "brute force" calculations and more on human-like analysis of position. This strategy can be summarized in the following way:

- in a given position perform a detailed tactical analysis by detecting relations between various groups of pieces (e.g. a rook is pinned by a bishop, a pawn is one move away from promotion line, the mobility of a knight is restricted to one possible move, a rook at the last line is a guard against mate, etc.),
- based on the above analysis define the list of various tactical and positional goals and select moves, which potentially lead to fulfilling these goals (such goals may include king's safety, pawn promotion, advantageous material exchange, strengthening the pawn structure, sacrifice of a (white) knight at f7, pins, forks, sacrifice of a (white) bishop at h7, double attacks, etc.),
- investigate these plausible moves prior to the other ones.

The general idea of selective, goal-oriented search proposed in the above strategy is also the basis of many current AI/CI game playing programs. Although Shannon did not provide the recipe for how to perform such detailed analysis and how to link the analytical results with subgoal definition process, the ideas presented in [296] are undoubtedly corner stones of several of today's approaches.

It is worth noting that although none of the three above algorithms (A, B, and modified B) involves any explicit learning, Shannon admitted that "Some thought has been given to designing a program which is self-improving, but although it appears to be possible, the methods thought of so far do not seem to be very practical. One possibility is to have a higher level program which changes the terms and coefficients involved in the evaluation function depending on the results of games the machine has played. Slam variations might be introduced in these terms and the values selected to give the greatest percentage of "wins."" These thoughts materialized only a few years later in Samuel's checkers playing program and in many other successful attempts afterwards.

Basic AI Methods and Tools

3.1 Definitions and Notation

In order to describe basic search algorithms used in classical AI framework some common notation needs to be introduced first. It will generally be assumed in this chapter that the evaluation function, i.e. the function returning the estimation of a goodness of a given game position is of linear form, represented as a weighted sum of real-valued (usually binary or integer) features. This is the most common situation, encountered in almost all advanced programs. More precisely, the top playing programs often use a set of linear evaluation functions, each of them devoted to particular phase of the game. For example Chinook [284] described in chapter 4.2 uses four such functions.

The task of defining the efficient evaluation function can be regarded in two perspectives: selection of the suboptimal set of atomic features that compose that function, and optimization of weights in their linear, or sometimes nonlinear, combination. In CI systems usually either the set of atomic features, or the weight coefficients, or both are obtained through the learning process. Typically, neural networks, evolutionary algorithms, or temporal difference methods are used for defining the set of weights. This issue is further discussed in chapter 7. In AI systems the set of features that compose the evaluation function is usually predefined by human experts or obtained *a priori* in other way. Hence, the remaining subtask is appropriate choice of weights coefficients for (each) evaluation function. The most popular AI approach is again the careful hand-crafting process. Other possibilities include the use of optimization techniques, such as least mean square approximation, hill-climbing, or other local or global optimization methods.

In the remainder of this chapter, when describing the selected AI game-related algorithms, it is assumed that the set of game features $x_1, \ldots, x_n$ used in the evaluation function is complete and optimized. A particular choice of weights $w_1, \ldots, w_n$ is also assumed (cf. equations (7.1)-(7.4)).

Two-player games are represented by a game tree. The initial (current) position is represented in a root of a tree, all positions that can be attained in a single move (performed by the current player) are the immediate successors

J. Mańdziuk: Knowledge-Free and Learning-Based Methods, SCI 276, pp. 15–39.
springerlink.com   © Springer-Verlag Berlin Heidelberg 2010

of the root node. Then for each such node all positions obtained in one move are its immediate successors, and so on. Edges of the tree represent moves, which lead from the parent nodes to the child nodes. Hence, making a move in a game is equivalent to traversing the branch of a tree representing that move - i.e. going one level down in a tree. The average number of legal moves in each game position is called a *branching factor*. The higher its value the broader the game tree.

Below in this chapter basic tree search methods are presented and briefly discussed. Since these methods have been well-known within AI game community for years as fundamental AI techniques used in game research, their description will be restricted to the most important aspects, leaving out the details available in several dedicated publications. In the following description of the algorithms the active side (the one associated with the root node) will be called *a player* and the other side, *an opponent*. The term (making) *a move* will be equivalent to performing a half-move (a ply) in classical game terminology, i.e. a move denotes moving of the one (currently active) side only.

3.2   Game Tree Searching

The state space in nontrivial games is usually composed of billions of billions elements. In chess, for example, with the average branching factor equal to about 35 and with the average game length of about 80 moves (40 moves on each side) there is around 35^{80} possible lines of play. Certainly these paths do not lead to pairwise different, unique positions, but nevertheless it is estimated that the number of possible, different states equals approximately 10^{50} [296]. Assuming that the assessment of a chess position requires 1 nanosecond (10^{-9} sec.) in average, the evaluation of all possible chess positions (exploration of the full game tree) would require about 3×10^{35} years - an enormously long time! Hence in computer game analysis several cutting-tree techniques are applied in order to decrease the complexity of a game tree search problem.

The easiest, though not generally recommended, method is restriction of the search depth to some predefined value, say d. In such a case all possible lines of play of length equal to d (or less in case a leaf node is encountered) from a given position are considered. Hence a program, in each position, searches in average d^w moves, where w is the average branching factor.

The obvious disadvantage of fixing the search depth limit is the impossibility to properly evaluate instable positions. A typical example is a multiple exchange of pieces which is "broken" by the depth d (i.e. starts before the search limit is achieved, but lasts beyond dth move). This disadvantage is called the *horizon effect* - the program cannot search beyond the horizon line defined by d. An immediate solution is application of selective search, which in case of instable positions is extended beyond depth d until a quiescent position is reached.

In the remaining part of this chapter some of the classical AI search algorithms are introduced, starting from minimax and alpha-beta, through

SCOUT and Negascout/PVS, up to MTD family of algorithms and UCT - the latest accomplishment in this area. Certainly these methods do not cover the whole list of search methods developed by AI community. Some widely-known achievements, e.g. SSS* [311] or B* [25] are not presented, since the objective of this chapter is an overview of selected ideas that will be later on utilized in the book, rather than in-depth presentation of the area.

3.2.1 Minimax Algorithm

In practically all game tree search algorithms the root node of the tree is associated with the current game position, i.e. a position in which a player from whose side the analysis is performed is to make a move. This node is typically denoted as a MAX node, since the player would like to maximize its evaluation score returned after analysis of the game tree. The next layer represents nodes associated with positions in which the opponent is to make a move. These nodes are the MIN nodes, since the opponent would choose continuations leading to the lowest possible score (from the player's perspective). The MAX and MIN layers alternate in a game tree until a predefined search depth d is reached (for the moment, the horizon effect is ignored and it is assumed that the search depth value is fixed). The $MIN - MAX$ notation is related to the classical tree search algorithm called *minimax* introduced in 1944 by von Neumann and Morgenstern [333] (based in part on von Neumann's theoretical investigations presented in [332]) in the version of full game tree expansion and enhanced by Shannon [296] who proposed using a *depth cutoff*. Once the tree is expanded to depth d, the leaf nodes are evaluated using some evaluation function. Usually an even value of d is used, i.e. the leaves belong to the MAX layer. In the next step the values of all preceding nodes in MIN layer are calculated as the minimum value of all their child's evaluations. Then the values for the previous MAX layer nodes are calculated as maximum over all their child's nodes, and so on, until the root node is reached. An example of the way the minimax operates and the pseudocode of the algorithm are presented in figures 3.1 and 3.2, respectively. Note that, in practical implementations, the minimax algorithm, should return not only the evaluation of a root node s, but also the first move on *the principal variation path*, i.e. a move to be made on the most promising path found. A detailed description of the algorithm can be found for example in [164, 276].

In practice, it is common to use the so-called *negamax* version of the algorithm, whose implementation is a bit simpler than original minimax. The idea consists in the observation that

$$\min\{x_1, x_2, \ldots, x_n\} = -\max\{-x_1, -x_2, \ldots, -x_n\} \qquad (3.1)$$

and in the fact that for the zero-sum games, in any game position the sum of minimax values for the player and his opponent is always equal to zero. In negamax notation node values are propagated to the upper level with *changed signs* and at each level (being either MAX or MIN) a branch with a *maximum*

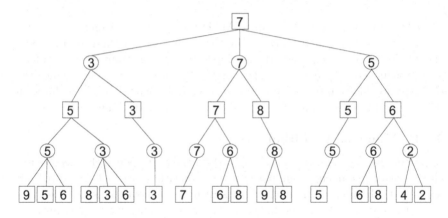

$\boxed{7}$ - MAX levels   ③ - MIN levels

Fig. 3.1. An example of the minimax search. The values represent nodes' assessments. The analysis starts from the lowermost layer and proceeds backwards up to the root node

```
function Minimax( s, d)
    if ((s == NOCHILD) or (d==0))
        return (evaluation(s));
    else
        if (s == MINnode) best_score:= +∞; /* an opponent is to play */
        else best_score:= −∞; /* a player is to move */
        foreach s' /* a child of s */
            score:=Minimax(s', d-1));
            if (s' == MINnode) best_score:=min(score, best_score);
            else best_score:=max(score, best_score);
        return (best_score);
```

Fig. 3.2. A pseudocode of minimax algorithm

value is selected. In negamax representation in the pseudocode presented in fig. 3.2 *best_score* value is always initialized with −∞ and the lines

if (s' == MINnode) best_score:=min(score, best_score);
else best_score:=max(score, best_score);

are replaced by

best_score:=max(-score, best_score);

3.2.2   Alpha-Beta Search

In many situations there is no need for full exploration of the tree. During the search some nodes may be omitted since their evaluation would not affect the final minimax result. Assume for example that a given node *s* of type MAX has two successors s_1 and s_2 and the evaluation of s_1 is completed and equal to $V(s_1)$ and the algorithm starts calculations for node s_2 and returns the value of its first successor s_{21} and that $V(s_1) > V(s_{21})$. In such a case further exploration of node s_2 is irrelevant, since s_2 is a MIN node and its value would be at most equal to $V(s_{21})$, which would anyway lose in comparison with $V(s_1)$ in the final competition in node s.

The above observation is the basis of the alpha-beta pruning method initially proposed in [46, 88] and fully developed and analyzed in [169], where the idea of deep cut-offs was introduced. The pseudocode of alpha-beta is presented in fig. 3.4. An example of its performance is depicted in fig. 3.3. A leaf node F is not explored, because the value of its parent node can be at most 3 (already known value of leaf E) and will be smaller than 5, which is the value of a competitive node in the MAX (third) layer. Similar reasoning can be applied to node J. Finally, the whole rightmost branch including the leaves $N - Q$ will not be searched, because the value of the parent node in the MIN layer cannot exceed 5 (a value propagated from the leaf M upwards) and as such will be not competitive to the value 7 of the middle branch.

More formally, for each currently evaluated node, the algorithm maintains two values α and β which represent the lower bound for the evaluation of a MIN node and the upper bound for the evaluation of a MAX node, respectively. If for any node s its momentary estimation falls out of its respective bound the whole subtree rooted at s is left out, since the position associated with s will not be played assuming perfect play of both sides. In its classical form the algorithm is initially run with $(\alpha, \beta) = (-\infty, +\infty)$ and this interval is gradually narrowed.

Analysis of fig. 3.4 leads to an observation that the largest cut-off can be achieved if in each node the child node which is analyzed first represents the best move. Using the convention of searching from left to right, in our example the middle branch should be located on the left. Also the order of the leaves should be optimized. One such optimal ordering of leaves is depicted in fig. 3.5. Observe that the main advantage comes from the appropriate ordering of the successors of s and the next generation nodes, since it allows for a substantial cut-off. In the case of fig. 3.5, initially the value of 7 is propagated from the first branch. Then in the second branch, due to choosing node G estimated as 3 (which is less than 7) as the first candidate move, the remaining part of a middle branch is omitted. Similar reasoning applies to the third branch.

For the constant branching factor w, computational complexity of minimax algorithm equals $O(w^d)$. The same is the pessimistic complexity of alpha-beta, since in the case of reverse-optimal ordering no cut-offs take place. The complexity of alpha-beta in the optimistic case (i.e. with optimal ordering of nodes) equals $O(w^{\frac{d}{2}})$, which - as proven in [254] - is the minimum number of nodes that need to be expanded in order to calculate the minimax value in a given position. This means that the optimal ordering of moves allows doubling the search depth within the same amount of time. The average asymptotic complexity of alpha-beta equals $O((\frac{w}{\log w})^d)$, for $w > 1000$ [169]. In the case of games with reasonable branching factors ($w << 1000$), the average algorithm's complexity equals $O(w^{\frac{3d}{4}})$. In practice, considering first the capture moves, followed by the threat moves, in most of the cases leads to fairly efficient ordering, not far from the optimal one.

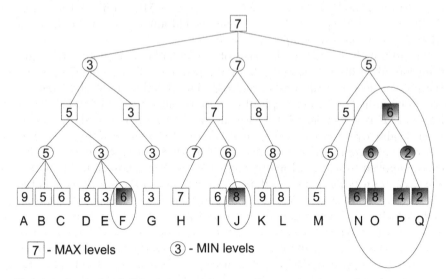

Fig. 3.3. Example of alpha-beta performance. Grey nodes denote the part of a tree that is not searched. See description within the text for details

Knuth and Moore [169] proved that alpha-beta yields the same results as minimax. Actually, the minimax can be implemented as alpha-beta with the infinite initial search window (see fig. 3.6).

A similar note concerning the negamax notation, which was placed in the previous section also applies to implementations of alpha-beta pruning algorithm. A pseudocode of the so-called fail-soft alpha-beta algorithm in negamax convention is presented in fig. 3.7.

3.2.3   Heuristic Enhancements

The efficiency of search algorithms depends on the number of nodes that need to be evaluated in the process of estimating the value of the root node s. In case of alpha-beta-based search methods, despite intrinsic differences between particular implementations, the common path for possible enhancement is appropriate sorting of the child nodes in each of the evaluated positions, from the most promising move to the least promising one. In the optimal situation, in each node the best continuation should be considered first. Certainly, sorting mechanisms need to be much less computationally expensive than the calculation of node's true minimax value, otherwise sorting the moves would be pointless.

In some of the search methods (e.g. in the MTD(f) family of algorithms) the information required for fast, heuristic node pre-ordering is acquired from previous runs of the algorithm.

```
function alphabeta( s, d, α, β)
    if ((s == NOCHILD) or (d==0))
        return (evaluation(s));
    else
        if (s == MINnode) /* an opponent is to play */
            foreach s′ /* a child of s */
                β:=min(β, alphabeta(s′, d-1, α, β));
                if α ≥ β return (α);
            return (β);
        else /* a player is to play */
            foreach s′ /* a child of s */
                α:=max(α, alphabeta(s′, d-1, α, β));
                if α ≥ β return(β);
            return (α);
```

Fig. 3.4. A pseudocode of alpha-beta algorithm

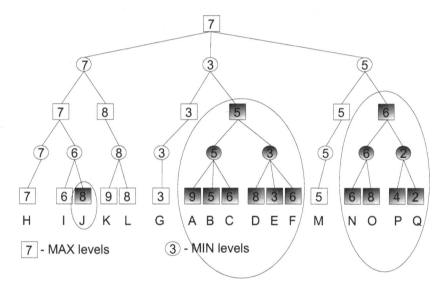

Fig. 3.5. One of the possible optimal orders of the nodes in alpha-beta algorithm assuming that the search proceeds from left to right. Grey nodes denote the part of a tree that is not searched. See description within the text for details

```
function Minimax(s, d)
    return (alphabeta(s, d, −∞, +∞))
```

Fig. 3.6. A pseudocode of minimax as a special case of alpha-beta

Killer Moves

One of the heuristics used for supporting the move ordering process is the *killer heuristic* [2], which assumes that moves which previously led to significant

```
function alphabetaNM(s, d, α, β)
    if ((s == NOCHILD) or (d==0))
        return (evaluation(s));
    else
        best_score:=α;
        while ((best_score < β) and (there are unvisited successors in position s));
            s':= next unvisited child of s;
            if (best_score > α) α:=best_score;
            score:= -alphabetaNM(s , d-1, −β, −α);
            best_score:=max(score, best_score);
        return (best_score);
```

Fig. 3.7. A pseudocode of negamax version of alpha-beta algorithm

cut-offs at a certain depth in other branches of the game tree are likely to produce a cut-off in the current (possibly similar) position. A certain number of such "killing" moves is maintained for each search depth and whenever the search reaches the respective depth these moves are considered first (if legal in the analyzed position). In the simplest replacement scheme, whenever a non-killing move produces a cut-off, it replaces one of the killing moves associated with the respective search depth.

History Heuristic

The *history heuristic* [281, 282] is a generalization of the killer heuristic, which maintains the rating of all *sufficient* moves found in previous searches. A sufficient move is one which either caused a cut-off or, if no cut-offs took place, yielded the best minimax value (note that these two characteristics are not equivalent). The idea lying beneath this heuristic is the same as that of killer heuristic: a move previously found to be a strong one is more likely to be efficient in an unknown position than a move "without history." In the simplest form the move is represented as a pair of squares *(from, to)* without differentiating between the pieces making the move. For example in chess a move can be represented as a 12-bit number [282], where the first six bits denote the square *from* and the other six - the destination square. Whenever a certain move is found to be *sufficient* in the sense defined above the respective value in a table of size 64×64 is incremented by a value 2^{d_t}, where d_t is a depth of the subtree searched from the node representing this move towards the leaves. This way sufficient moves close to the root are weighted higher than the ones close to the leaves of the searched tree. In other words, the deeper the subtree searched, the higher the confidence in the quality of a chosen move. History heuristic is a very effective and low cost technique of narrowing the size of the searched tree. It is used in majority of world-class AI board game playing programs, commonly together with transposition tables described below.

Transposition Tables

Transposition table is one of the most frequently used heuristic search enhancements in game playing programs. The name comes from the observation that usually a game position (equivalently: a node in a game tree) can be reached by various sequences of moves differing by the order in which they are executed. For example, in chess opening, the sequence

$1.d2 - d4, g7 - g6, 2.c2 - c4, d7 - d6$

leads to exactly the same position as the *transposed sequences*:

$1.d2 - d4, d7 - d6, 2.c2 - c4, g7 - g6$ or
$1.c2 - c4, g7 - g6, 2.d2 - d4, d7 - d6$ or
$1.c2 - c4, d7 - d6, 2.d2 - d4, g7 - g6.$

Transposition table is used for temporal storing of assessments of searched nodes. It is usually implemented as a hashing table, e.g. using the Zobrist hashing [353, 354]. Based on the representation of state s, a hashing index is created, which is used for accessing the appropriate element of the table when storing or retrieving the value of s. Usually, information stored for state s includes the best move found for that position (i.e. the one which caused cut-off or had the best minimax score) with its minimax value and the search depth used to calculate this move. Additionally the number of nodes searched in a subtree rooted at s can be stored. Whenever the search enters the state s (possibly in another branch of a game tree) the respective table entry is retrieved (in constant time, without searching the table) and if not empty, a decision is made whether the depth of previous search is adequate. If so, the stored best move and its minimax evaluation are used by the program without the need for search repetition. If the answer is negative, the stored value is used as the first approximation of the node's assessment and the best move found in previous searches is the first one to be searched in the current, deeper search. In effect the amount of calculations required in node s is reduced.

Due to memory limitations, the size of transposition table is usually far smaller than the number of all possible moves in all different positions. Hence, at some point collisions in accessing the table may occur. There are generally two types of such collisions: either two different positions generate the same key or two different positions generate different keys, which point to the same element of the table. The latter situation may happen if generated keys exceed the size of the array and after dividing them modulo size of the table the resulting values are equal. Both situations are undesirable and require the use of appropriate replacement scheme [44, 179]. The most popular policies are: *deep* - replace if the search depth of a new position is not less than the search depth of currently stored position, *big* - replace if the number of nodes in a searched subtree in a new position is greater or equal than the respective number in a stored position, and *new* - always replace the current entry with a new one.

Experimental analysis of the above replacement schemes [44, 179] favors the *big* replacement scheme, before *deep*, with *new* being the least effective.

On the other hand, along with increase in the size of transposition table, differences between particular schemes tend to become negligible. In the author's experiments with give-away checkers and real-value evaluation function [213], the replacement policy *new* appeared to be the most effective for the transposition table of size 2^{20}.

Yet another possibility is to combine two of the above schemes by maintaining two tables: *deep* and *new* - the so called *two_deep* replacement scheme. If a collision occurs, it is first checked if the *deep* scheme is applicable to the first table. If so, replacement in the *deep* table is made and the value from the *deep* table is moved to the other, auxiliary table. If not, replacement takes place in the *new* table. This way the first table always stores the entries related to the highest searched trees for the stored game positions. In the minimax evaluation search both tables are searched for the possible cut-off moves, and if none is found, the best moves from one/both of them are used as the first approach.

In summary, transposition tables are commonly used in top playing programs allowing substantial reduction of computational load by avoiding search repetition. They are especially efficient in the task of preliminary move ordering and narrowing the range of minimax alpha-beta search, e.g. in the MTD(f) algorithms described below in this chapter.

Refutation Tables

Refutation tables [2] are used to store the sequences of moves from the root s to the leaves. Besides the principal variation path, other, inferior lines of play are stored in these tables together with either the correct minimax value or its upper bound estimation. Refutation tables require less memory than transposition tables and were preferred in early years of game AI programming mainly due to the severe memory limitations at that times. Nowadays they are usually used in conjunction with the iterative deepening technique described below, assuming that paths obtained for d-ply search are indicative (good or wrong) also for depth $d+1$. The moves that proved to be efficient or the ones refuted in previous search are likely to have similar characteristics in one ply deeper search. Refutation tables can also be used for moves ordering.

Iterative Deepening

Iterative deepening [132, 217] consists in repeatedly running the alpha-beta search for gradually increasing depths from $d = 1$ to $d = n$, for some n. The method is commonly used when the search is restricted by time constraints, for example in tournament situation. The main advantage of this type of search is the possibility to stop the process at any time, say for $d = i, 1 < i \leq n$ and return the best move found in the preceding search phase (usually with $d = i-1$ or $d = i-2$, depending on the algorithm's design). This way the time allotted for a given move can be effectively used, without essentially knowing the greatest achievable depth of search. In the case of classical alpha-beta approach with some predefined depth n, if the search is not completed, the resulting move may be far inferior to the optimal choice, due to the depth-first-search nature of the algorithm.

In practice, the iterative deepening procedure stops either when search at $d = n$, for some predefined n, is completed or the time allotted for the search is fully utilized.

Complexity of the iterative deepening may, at first sight, seem high, since the upper nodes in the tree are searched repeatedly several times. However, in regular game trees, with approximately stable branching factor, the leaves of the tree account for the most of the overall search burden. More precisely, in typical game tree the complexity of the iterative deepening procedure is of the same order as the complexity of its last phase, with $d = n$. Hence, from the computational complexity viewpoint repeatable searching through the higher nodes is of negligible cost - especially if the average branching factor w is high. Additionally, transposition and refutation tables as well as history or killer heuristics can be used at each search phase, allowing for substantial time savings by reusing the information acquired in previous phases while searching at greater depths.

Since in the iterative deepening procedure the leaves at level n are visited once, the nodes at depth $d = n - 1$ are expanded twice, ..., and the root node is expanded $n + 1$ times, assuming the tree is ideally balanced and has the branching factor w, the total number of expansions D is equal to [276]:

$$D = \sum_{i=1}^{n+1} i \cdot w^{n+1-i} \tag{3.2}$$

In chess, for example, with $w = 35$ and $n = 10$, the number of expansions equals approximately $2,92E + 15$ compared to $2,76E + 15$ of a one pass search performed with depth limit $d = n$. Complexity overhead of iterative deepening in this case is lesser than 6%!

Null Moves

Null move heuristic [23, 148] is used in chess programs and aims at speeding-up the alpha-beta search by guessing the cut-off moves without performing deep, full-width search of a subtree. The idea is based on the observation that in majority of chess positions making a move causes the game position to improve (from the moving player's perspective). Hence, if the player whose turn it is can refuse the right to make a move and still have a position strong enough to produce a cut-off, then this position would almost certainly be a cut-off if the player actually made a move.

In practical implementations, in order to computationally benefit from applying this heuristic, after (virtually) forfeiting the turn of the player, an alpha-beta search is performed to shallower depth than it would have been in case a move had been performed by the player. Due to shallower search the cut-off is found faster and with lower computational costs. If cut-off is not produced, regular, full-depth search has to be performed. Certainly, the shallower the search the less reliable the minimax estimations, so in practice the null move search is usually no more than $2 - 3$ ply shallower than the regular search. In typical chess positions the frequency of proving cut-offs (the successful null move searches) is high enough to justify the use of the method.

There are, however, game positions in which null move heuristic should not be employed since in these situations the method would almost universally cause the wrong judgment. These positions are commonly classified as *zugzwang* positions (from German *forced to move*). In case of zugzwang the choice the player has is among bad moves only and actually they would be better off if forfeiting the move was allowed in chess. Application of null move heuristic may therefore produce a wrong cut-off, which would not have taken place, had the regular alpha-beta search been employed. In order to avoid the above, null move heuristic should not be used in positions in which the player is in check or has few pieces remaining - special care should be taken in positions when the player is left exclusively with king and pawns. It is also strongly recommended that no more than one null move be considered on each explored game trajectory.

Selective Extension of the Search Depth

The most popular method of gradual extension of the search depth is *iterative deepening* described above in which the search depth is increased by an integer increment, usually one or two ply at a time. This is a powerful, low-cost and safe technique allowing for breaking the search and returning the best solution found in the previously completed search iterations at any point in time. The characteristic feature of iterative deepening is that it works in a uniform manner, gradually extending the depth of the whole searched tree. In majority of popular games, however, there is a need to extend the search along specific paths only, without the necessity of increasing the search horizon in the whole tree. These situations include non-quiescent positions, for example in the middle of a sequence of material exchanges or after checking or check evasion moves (in chess) or generally after the moves causing immediate and serious threats, which need to be resolved before backing up the position value in the game tree. Such situations require selective extensions of particular search paths.

Several methods have been developed in this area - most of them relying on the domain specific knowledge allowing for the estimation of the "forcedness" of a given move. These methods are commonly referred to as *selective extensions* or *selective deepening methods*. The common ground for all of them is the use of static measures for deciding whether the move (position) requires deeper exploration. The main advantage of using static measures when estimating the forcedness of a move is their simplicity, since such assessment is based on the enumerative catalogue of different features, which are generally easily verified in a given position. On the other hand, in more complicated games it is difficult to provide a complete knowledge-based set of features which would cover all possible game situations. Additionally, such a set of static features does not take into account the dynamic aspects of the position (e.g. tempo, tactical sacrifices, long term strategic goals, etc.).

One of the interesting alternatives to static selective extensions is the method called *singular extensions* proposed by Anantharaman, Campbell and Hsu [7] which uses dynamic measures when assessing the forcedness of a move and where the important lines of play are identified in a dynamic and game-independent manner. In short, a move is considered as *singular* if its estimated value is significantly better than the values returned by all its siblings. The notion "significantly better" is expressed numerically as a threshold of the difference between this singular move and the comparative alternatives.

Another dynamic, game independent approach to the problem of variable depth search are *conspiracy numbers* proposed by McAllester [221]. In a given game tree each conspiracy number C_n, for integer n, represents the number of leaf nodes that must change their scores (as a result of a deeper search) in order to change the score in the root node to become equal to n (these nodes *conspire* to make that change happen). The algorithm uses a threshold defining the feasibility of potential change in the root node. If for any n, C_n exceeds that threshold, changing the root node's value to n is considered unlikely and is not of interest anymore. The algorithm terminates if there is only one value n left such that C_n is below the threshold.

In order to reach the above terminal state the algorithm selectively chooses the leaf nodes for further expansion in order to narrow the range of possible minimax estimations in the root node. The method does not guarantee theoretically that a best move will be found, but the confidence in the algorithm's outcome grows with increase of the threshold - see [221, 283] for the details and discussion on the implementation issues.

3.2.4   SCOUT and Negascout (PVS)

SCOUT

In 1980 Judea Pearl published [247] a new tree search algorithm called SCOUT, which, he claimed, might be superior to alpha-beta in case of game trees with high height-to-branching-factor ratio. The algorithm is based on the following idea: first one branch of the tree is searched and value of the leaf node is assessed according to a given evaluation function. This value is assumed to be the true minimax value of a tree, i.e. a true estimate for its root node s. The objective of the algorithm is to verify this assumption by exploration of the nodes belonging to the searched tree. Implementation of this general idea requires two functions, *EVAL* and *TEST*.

Procedure TEST$(s, v, >)$ is used to check whether given node s satisfies the condition $V(s) > v$, where $V(s)$ denotes the minimax value of node s. Its implementation is based on the same observation that led to definition of alpha-beta algorithm. Namely, in case of MAX node the procedure should return TRUE as soon as a descendant of s with value greater than v is found; conversely, in case of MIN node, it should return FALSE as soon as it identifies a successor of s with value less than v. The analysis of minimax values of descendant nodes is performed by a recursive call to TEST procedure for each of them.

Procedure EVAL(s) is responsible for finding the actual minimax value of node s. If s is a MAX node, this function first invokes itself recursively on s_1 - the first successor of s - and assumes that $V(s) = V(s_1)$. Next, TEST procedure is called for each child node s_i to validate assumption that $V(s_i) \leq V(s)$. If in the process for any i it is found that $V(s_i) > V(s)$, then full calculation of $V(s_i)$ is performed by calling EVAL function. $V(s)$ is assumed from now on to equal this new value and it is used in calls to TEST procedure for all subsequent nodes. MIN node analysis is performed analogically with the obvious change of inequality operators.

Although SCOUT algorithm may have to evaluate some of the branches of the game tree twice, it can be mathematically proven that it still achieves the optimal branching factor (just like alpha-beta). Moreover, many calls to TEST result in performing a cut-off, thus reducing the search complexity, as they are much faster than full alpha-beta searches. This fact causes SCOUT algorithm to outperform alpha-beta at least in case of some game trees.

Negascout

In 1982 and 1983 two very similar algorithms were developed: Tony Marsland and Murray Campbell's *Principal Variation Search* (PVS) [219] and Alexander Reinefeld's *Negascout* [262], which successfully incorporated the SCOUT algorithm's concepts into an alpha-beta in its typical negamax notation. The latter will be described in more detail in the following paragraphs.

Negascout searches the game tree in a way similar to SCOUT, but makes use of negamax notation and introduces some optimizations - most notably always traverses the tree with two bounds instead of one-side constraint of TEST function in SCOUT. Its implementation, assuming integer values of the evaluation function, is presented in fig. 3.8. It is based on a negamax formulation of alpha-beta algorithm, introducing only a few new statements.

Just like in regular alpha-beta, static evaluation function's return value is used as a minimax value of leaf nodes. In case of regular nodes, their first

```
function Negascout(s, α, β, depth)
1:  if (depth == MAX_DEPTH) then return evaluate(s);
2:  m := -∞;
3:  n := β;
4:  for each {successor sᵢ of s}
5:      t := -Negascout(sᵢ, -n, -max(α, m), depth + 1);
6:      if (t > m) then
7:          if (n == β or depth >= MAX_DEPTH - 2) then m := t;
8:          else m := -Negascout(sᵢ, -β, -t, depth + 1);
9:      if (m ≥ β) then return m;
10:     n = max(α, m) + 1;
11: return m;
```

Fig. 3.8. A pseudocode of Negascout algorithm

successor is evaluated with the regular window search of $(-\beta, -\alpha)$. Negation of its value is assumed to be the minimax value of node s and assigned to variable m. Negascout then "scouts" subsequent successors of node s from left to right using a zero-width window $(-m - 1, -m)$. Assuming integer evaluation values, this search is bound to fail, but its failure direction (high or low) forms basis for decision whether full evaluation of the node is required.

Whenever the zero-width window search fails low, it signals that the minimax value of the tested node is lower than the current maximum value of its siblings and, as such, will not influence its parent's value. Algorithm may, therefore, safely continue scouting subsequent nodes. In general, when the null window search fails high, Negascout has to revisit the same subtree with wider search window. Only in case of searches with depth lower or equal 2 is this research unnecessary as an exact minimax value is always returned. It is worth noticing that it is not required to research with the full window $(-\beta, -\max(\alpha, m))$ as would have been done in regular alpha-beta algorithm, but it suffices to use narrower window $(-\beta, -t)$.

In practical implementations of Negascout algorithm, a path to the leftmost leaf causing a failure should be stored, so that it can be followed first during a potential revisit of the subtree. Additionally, identification of principal variation of the game tree turns out to be a side-effect of this feature. Negascout proves to be more effective than traditional alpha-beta pruning for many game trees with high branching factors and depths. Negascout also profits significantly from most alpha-beta algorithm enhancements, including above all transposition tables.

3.2.5   MTD(f) and MTD-bi

Integer Evaluation Values: MTD(f) Algorithm

MTD(f) [253, 255] algorithm takes Pearl's TEST procedure idea to the extreme. It is nowadays considered to be the fastest game tree search method. Its name is an acronym for *Memory Enhanced Test Driver* and it belongs to a class of MTD algorithms, which all involve making repeated calls to MT function. MT stands for *Memory Test* and denotes a null-window Alpha-BetaWithMemory search - a procedure based on SCOUT's TEST function (and its subsequent reformulation as part of Negascout).

A pseudocode for MTD(f) is represented as figure 3.9(a). As explained previously, the algorithm consists of multiple passes of a zero-window alpha-beta search. Each pass yields a high or low bound on the evaluated minimax value, depending on whether the search fails, respectively, low or high. The algorithm stops when those bounds converge to the same value. Since all the subtree searches are performed with a zero-width window, cut-offs occur very frequently. This property and the usage of memory-enhanced version of alpha-beta procedure conspire to high efficiency of the algorithm. Value f - an argument of MTD(f) function - represents the initial approximation of the minimax value being calculated. Previous MTD(f) analysis result is used as this value in all MTD(f) calls but the first one – since no previous results are available then, $f = 0$ is usually assumed. The quality of initial approximation of f may significantly influence the algorithm's performance, as searches with better initial value of f require fewer alpha-beta passes.

For clarification, figure 3.10 contains a pseudocode of the most popular version of AlphaBetaWithMemory routine referenced in MTD-class algorithms. It makes use of transposition tables (described in section 3.2.3).

Floating Point Evaluation Values: MTD-bi Algorithm

MTD-bi algorithm [255] is another MTD variation, assuming continuous evaluation values and making use of bisection method (see figure 3.9(b)). MT searches are performed with a window $(b - \epsilon, b)$, where b defines alpha-beta window position and ϵ is a small, constant, arbitrarily set window's width. Analogically to MTD(f) algorithm, whenever the result of MT procedure g is lesser than $b - \epsilon$, g becomes the new upper bound f^+ and if $g > b$, then it becomes the new lower bound f^-. In case of MTD-bi, the alpha-beta does not, however, necessarily always fail. MT return value of $g \in (b - \epsilon, b)$ signals that the exact minimax value has been found. Until this condition if fulfilled, alpha-beta searches are performed repeatedly with null-windows centered around $(f^+ + f^-)/2$.

In [212] MTD-bi algorithm was proven to be more efficient in the domain of floating point values – i.e. in cases when minimax evaluation values belong to continuous interval of real numbers. Explanation involves the fact that, in case of MTD(f), whenever $g = b + \delta, \delta > 0$, the new value of b would equal $b + \delta + \epsilon$. Since ϵ is by definition close to zero, in case of small δ, the change in the value of b would be very small, as well. At the same time, small value of δ does not necessarily coincide with b being close to sought minimax value. The effect is that MTD(f) might require many steps to finally reach the minimax value of the analyzed node. The use of bisection-based calculation of b values successfully remedies this problem.

function MTD(f)
1:   $f^+ := MAX; f^- := MIN;$
2:   **if** ($f = MIN$) **then** $b := f + 1$ **else** $b := f;$
3:   **repeat**
4:       $g := MT(s, b - \varepsilon, b);$
5:       **if**($g < b$) **then**
6:           $f^+ := g;$
7:           $b := g;$
8:       **else**
9:           $f^- := g;$
10:          $b := g + \varepsilon;$
11 **until** $f^- \geq f^+$
12: **return** $g;$

(a) MTD(f) algorithm.

function MTD-bi
1:   $f^+ := MAX; f^- := MIN;$
2:   **repeat**
3:       $b := (f^+ + f^-)/2 + \frac{\varepsilon}{2};$
4:       $g := MT(s, b - \varepsilon, b);$
5:       **if** ($b - \varepsilon < g < b$) **then** **return** $g;$
6:       **if** ($g < b$) **then** $f^+ := g$ **else** $f^- := g;$
7:   **until** forever

(b) MTD-bi algorithm.

Fig. 3.9. MTD(f) and MTD-bi algorithms

function alpha-beta_with_memory(s, d, α, β)
    if (TT(s).retrieve == OK)
        if (TT(s).lowerbound $\geq \beta$) return (TT(s).lowerbound);
        if (TT(s).upperbound $\leq \alpha$) return (TT(s).upperbound);
        α := max(α, TT(s).lowerbound);
        β := min(β, TT(s).upperbound);
    if (d == 0) g := evaluate(s);
    else if (s == MAXnode)
        g := $-\infty$; a := α; c := firstchild(s);
        while ((g < β) and (c != NOCHILD))
            g := max(g, alpha-beta_with_memory(c, d-1, a, β));
            a := max(a, g);
            c := nextbrother(c);
    else if (s == MINnode)
        g := $+\infty$; b := β; c := firstchild(s);
        while ((g > α) and (c != NOCHILD))
            g := min(g, alpha-beta_with_memory(c, d-1, α, b));
            b := min(b, g);
            c := nextbrother(c);
    if (g $\leq \alpha$) TT(s).upperbound := g;
    if ((g > α) and (g < β))
        TT(s).lowerbound := g;
        TT(s).upperbound := g;
    if (g $\geq \beta$) TT(s).lowerbound := g;
    return (g);

Fig. 3.10. A pseudocode of AlphaBetaWithMemory algorithm

3.3 Rollout Simulations and Selective Sampling

Rollout simulations or *rollouts* is a very interesting technique of estimating individual move (position) strength based on the results of multiple Monte Carlo simulations. The idea proved to be especially useful in imperfect-information games such as backgammon, Scrabble, poker, or bridge. The common characteristic of these games is very high branching factor, which is the result of either hidden information or the element of chance. For these games employing traditional full-width search is infeasible and Monte Carlo sampling method represents a plausible alternative.

For a given candidate position (game situation) thousands of games are played with random choice of the hidden information (opponent's cards in poker or bridge, or tiles in Scrabble) or the random dies rolls (in backgammon). The average result of such multiple simulations provides the expected empirical strength of considered position. The idea is fairly simple and surprisingly powerful, given its very general formulation and universal, game independent applicability.

Strictly speaking, the above Monte Carlo-type sampling based on uniform distribution of the unknown data is mainly used in the *off-line* simulations aiming at proving the strength or weakness of a certain hidden data configuration (e.g. private cards in poker or bridge) or a certain move (e.g. the opening moves in backgammon). In the in-game situations performing full-range sampling simulations is infeasible and in such cases the *selective sampling* technique is used, in which particular moves or card distributions are selected according to some probability distribution and the higher the chance of occurrence of a given game situation (game position or card configuration) the statistically more frequently the situation is chosen in the sampling procedure. Moreover, the results of previously completed sampling simulations are used as a guidance for the prospective choices in the selective sampling process.

In both rollout and selective sampling simulations one of the concerns is whether simulations performed by a computer program, which is not a perfect player, are not biased by its imperfection or its "computer-style" playing. In other words the question that arises is whether the results of simulations made by imperfect player are to be trusted? Although no theoretical evidence proving the correctness of this approach is given, the experimental results speak for themselves. It seems appropriate to say that if only the program is a strong player it is generally sufficient to produce a trustworthy rollout estimations. One of the heuristic supports of this claim is the observation that in the simulation process potential fallibility of the program has similar impact on both playing sides and in effect is canceled out.

Computer rollouts proved to be extremely powerful and trustworthy, and are now widely applied in imperfect-information games. A summary of the developments in this area is sketched in the remainder of this chapter.

Backgammon

Application of simulations in game domain was proposed for the first time by Tesauro in his TD-Gammon program. Initially Tesauro used computer rollouts for *off-line* verification of the strength of particular moves by calculating the average over many possible game trajectories [319]. These off-line simulations proved the efficiency of certain moves, in particular in the opening phase of a game, leading to tremendous changes in human positional judgement in backgammon and to invention of entirely new, previously unknown openings, which currently are widely played by the top players.

In subsequent papers [321, 322] *on-line* simulations were discussed. The authors considered using either full-depth rollouts, in which game trajectories are expanded until the end-of-game position is reached, or using shallower simulations truncated at a few ply depth. Due to time requirements, the former variant could be applicable in case of simple linear evaluation functions, the latter would be recommended when nonlinear evaluation functions were applied.

Based on the experimental results the authors confirmed the usefulness of rollout simulations in online policy improvement of backgammon playing program. The advantage was observed for both linear function and the nonlinear Multilayer Perceptron (MLP) neural network representation (described in detail in section 5.1.1). On the implementation side, the simulation process could be fully parallelized since in case of Monte Carlo rollouts the sampled trajectories are pairwise independent.

In order to decrease computational load of the method, the authors recommended monitoring of the simulation process and, basing on preliminary simulation results, discarding moves (trajectories) which were very unlikely (according to some predefined measure of confidence) to become selected as the best ones. On the same basis moves which were evaluated as having similar strength to the actual best selection would be skipped in further simulations, since they might be considered equivalent choices [322].

Scrabble

The idea of selective sampling was also considered by Brian Sheppard in his Scrabble playing program MAVEN [298]. Similarly to Tesauro and Galperin's proposals, monitoring of the simulation process was applied in order to reject unfavorable lines of play already in the preliminary phase of the simulations. Furthermore, situations in which two moves could have been played in the same spot and with the use of the same tiles were regarded as "undistinguishable" and one of them was pruned out. Application of these two ideas caused substantial reduction in simulation time, approximately by a factor of eight, compared to "blind rollouts".

Poker

Simulations were also extensively used in several poker programs authored by Darse Billings, Jonathan Schaeffer and their colleagues from the University of Alberta in Canada. One of the programs, named *Poki*, described in detail in the literature [31, 32] takes advantage of Monte Carlo rollouts in the so-called pre-flop phase in Texas Hold'em[1] and also uses selective sampling in its betting strategy.

In the pre-flop phase each player receives his two private (hole) cards and, except for the actions taken by the preceding players, nothing more is known about the current game, yet. Certainly there are $\binom{52}{2} = 1326$ possible two-card choices and the potential value of each of them has been verified by Poki in extensive, several million trial-based, off-line simulations, yielding the initial approximation of the prospective hand's strength.

Betting decisions are supported by online simulations, which consist in playing the game many times from the current state till the end. Each such

---

[1] Texas Hold'em is the most popular variant of competitive poker played at tournaments and casinos all over the world. In Texas Hold'em the game is divided into five phases: *pre-flop* (when only private cards are dealt), *flop* (when the three community cards are revealed), *turn* and *river* (the fourth and the fifth common cards are dealt, respectively) and the *showdown* phase, which concludes the game. The rules of the game are described for example in [31] or in the basic poker compendiums written by Sklansky [304, 305].

trial is simulated twice: the first time to consider the *call* action, and the second time to verify the results of *raise* decision. In each such simulated scenario hidden cards are assigned to each of the players according to some probability distribution maintained by the program. These probabilities reflect the analysis of past opponent's actions (in previous rounds) and the actions undertaken in the current game. Similarly, the future community cards are sampled in the simulations according to some underlying probability distribution. This way the more probable hands and the more probable future community card selections are chosen more frequently than the less probable configurations, although even a very unlikely situation still has a chance to be drawn and tested in the simulations. The game is virtually played by simulating each player's betting decisions, which are strongly correlated with the choice of private cards the player possesses, the model of the player's behavior maintained by the program (conservative, bluffing, optimistic, ...) and the modeled betting strategy. Considering the above factors provides the likelihoods of the player's actions: *call*, *raise* or *fold*. Betting decision is selected at random consistently with these estimated likelihoods. The simulated game proceeds according to the above rules until the showdown, when the final result of the game, in terms of earning or losing money, is calculated.

The above simulations are repeated many times and for initial *call/raise* choices the averages of several respective trials produce the estimates of the consequences of both betting decisions. These values, together with the estimate of *fold* action, which is equal to zero since no future gain/loss is possible when folding the game, finally support the selection of betting decision [31].

Bridge

In bridge the idea of sampling simulations has been floating around since 1989 Levy's paper [189], but the first implementation in a tournament class program was realized by Ginsberg in his famous Ginsberg's Intelligent Bridge Player (GIB) [134, 135]. Monte Carlo simulations are used in GIB in both bidding and cardplay phases. In the latter, for each playing decision (choice of a card to be played) some number of simulations is performed according to the following scenario: a set of deals D consistent with the bidding information and the hitherto cardplay phase course is defined; then for each possible card selection and each deal from D the respective Double Dummy Bridge Problem[2] is solved; the move for which the sum of payoffs over all deals from D is maximal is finally selected to be executed in the actual play.

A similar procedure is run at each turn of the bidding phase. A set of deals D consistent with previous bids is first generated. Then, for each

---

[2] The Double Dummy Bridge Problem consists in answering the question of how many tricks are to be taken by each pair of players with all cards being revealed, assuming a perfect play of all sides. The ability to solve the DDBP perfectly is used in GIB to estimate the strengths of hands for a given (partly unknown) distribution of cards in a real game situation. Application of neural networks to solving the DDBP is discussed in chapter 8.4.

candidate bid b and for each deal from set D simulations of future auctions are performed assuming a bid b has been made. Future bids are predicted based on a specially pre-prepared database containing appropriate bids under various game circumstances. The database includes about 3 000 bidding rules. In the next step the DDBP related to the (virtually) declared contract is solved. The final bidding decision is the one which maximizes the result over all deals from D. Selective sampling simulations made by GIB were one of the major factors contributing to its high playing profile. GIB became the first bridge program capable of competing with expert human players, and the world's strongest machine player in 1998 and 1999.

Go

Due to a large branching factor, which prevents efficient use of full-width alpha-beta minimax search, Go is one of the natural target domains for rollout simulations. Go is a perfect-information game with no hidden data or stochastic factors, so playing the game deterministically multiple times would mean repeating the same game all around. The randomness is introduced to the simulations by the move selection policy used in the simulated games. In the simplest approach the rollout games are played from a given position up to the end of the game in a completely random manner[3]. An advantage of such random simulation scheme is that it does not require a sophisticated position evaluation function. Since the games are simulated till the end-of-game positions, all that is necessary is the ability to assess the final result (win/loss/draw) and the magnitude of the victory/loss. In majority of games assessing a final position is a significantly easier task than evaluation of a mid-game position. This is particularly true for Go in which, due to "dynamic nature" of the game, one of the main impediments in successful computer playing is difficulty in constructing efficient static evaluation function for position assessment.

The idea of purely random Monte Carlo simulations has one important drawback. The random choice of responses does not guarantee that the move which proved to be excellent in simulations made hitherto would not turn out to be poor in a particular game line (opponent's response), not yet simulated.

In order to lessen the impact of pure randomness and due to limited computational resources available, random simulations are often supported by some domain knowledge and/or shallow, indicative tree search to guide the selection process. In Go this idea was introduced by Brügmann in his *Gobble* program [47]. Brügmann proposed that the choice of moves in the simulations be related to their current evaluation, with higher probabilities of selecting moves already estimated to be strong. The estimation of a move's strength was calculated as the average outcome of all the games, in which this move was chosen to be played. This experience-based greedy strategy would have a

---

[3] This idea, in the domain of perfect-information games, was initially proposed by Abramson [1] in the 6×6 Othello.

tendency to settle in a local maximum and was therefore combined with the stochastic move selection, which was independent of previous move's performance. Brügmann proposed using a simulated annealing procedure to control the amount of randomness in the Gobble's selection process. At the beginning of the game random selection was dominant. Along with the game's progress performance-based evaluations of moves gradually prevailed in the selection.

The idea of using Monte Carlo simulations in high performance Go playing program was exploited by Bruno Bouzy in *Olga* [39, 43] and *Indigo*, and by Rémi Coulom in *CrazyStone* [72]. Indigo associates Monte Carlo simulations with Go domain knowledge [38].

CrazyStone efficiently combines the minimax tree search with Monte Carlo simulations by applying adaptive move selection procedure. If μ_i, σ_i^2 denote respectively the move's value estimated in previous simulations and the variance of this estimation, and assuming that the moves are ordered according to their respective μ values, i.e. $\mu_0 > \mu_1 > \ldots > \mu_n$, then the ith move is chosen with probability, which is proportional to:

$$u_i = \exp\left(-2.4\frac{\mu_0 - \mu_i}{\sqrt{2(\sigma_0^2 + \sigma_i^2)}}\right) + \varepsilon_i \qquad (3.3)$$

where ε_i is a special constant (dependent on i and n) which prevents u_i from approaching zero too closely. Formula (3.3) approximates the importance of a given move under the assumption of Gaussian distribution of its simulation-based estimations. It was applied in CrazyStone with remarkable successes - the program won the gold medal at the 11th Computer Olympiad on 9×9 board and silver medal at the 12th Computer Olympiad on 19×19 board.

Quite a different approach has been proposed in one of the recent papers by Bouzy and Chaslot who verified the possibility of applying reinforcement learning techniques to computing the urgencies of domain-specific patterns from the collection of predefined pattern templates and using this knowledge to guide Monte Carlo simulations [42].

Recently, a new development in Monte Carlo Go area - the UCT algorithm - has been announced.

3.3.1   UCT Algorithm

The idea of applying Monte Carlo sampling has been recently utilized in the state-of-the-art Go programs - Fuego [92] and MoGo [125, 126, 340], leading both programs to outstanding accomplishments: outplaying human Go professional players (9th dan and 5th dan, respectively), for the first time in the history without handicap, so far on a 9×9 goban (Go board). MoGo uses the UCT (*Upper bound Confidence for Tree*) search algorithm, proposed by Kocsis and Szepesvari [170] and adapted for the game tree search by Sylvain Gelly and his collaborators.

The UCT algorithm is a modification of UCB1 method [12] proposed in the multi-armed bandit framework. The K-armed bandit problem is a machine learning problem, in which a collection of K traditional slot machines (arms of a bandit) is considered. Each of the K arms (machines), when played, draws a reward according to some distribution, and these distributions are pairwise independent between the machines. The goal of the player is to maximize its collective reward gained through several iterative plays, where each play is equivalent to making the choice of a machine. In iteration n the choice is made based on the observations of rewards made in previous plays from 1 to $n-1$. More precisely, each machine $X_i, 1 \le i \le K$, yields a reward $X_{i,s}$ when played the sth time and $X_{i,s}, X_{j,t}$ are pairwise independent for $(i,s) \ne (j,t)$. Usually it is also assumed that all distributions $X_i = \{X_{i,1}, X_{i,2}, \ldots\}, 1 \le i \le K$ are defined according to the same law, but with unknown expectations $\mu_i, 1 \le i \le K$.

The optimal choice of an arm to be played is related to the so-called *exploration-vs.-exploitation dilemma*, which is regarded as a tradeoff between exploiting already known effective (rewarding) actions versus exploration of the new possibilities in the quest to find more rewarding actions. Certainly none of the two extremes is advisable: exclusive exploitation of already known actions would prevent the agent from searching the state space, whereas the policy based on exploration only would lead to a random search in the solution space without taking advantage of already found rewarding actions. Further discussion on this issue in the context of temporal difference learning in game domain is presented in section 9.1.4.

For the K-armed bandit problem, assuming that the rewards belong to the interval $[0, 1]$, it is recommended [12] to initially play each machine once (in any order) and then, if $T_{i(n)}$ denotes the number of times that machine i was played in the first n iterations, choose in the $n+1$th iteration the machine j that maximizes:

$$\overline{X}_j + \sqrt{\frac{2\log n}{T_j(n)}} \tag{3.4}$$

where

$$\overline{X}_j = \overline{X}_{j,T_j(n)} \quad \text{and} \quad \overline{X}_{j,m} = \frac{1}{m}\sum_{k=1}^{m} X_{j,k} \tag{3.5}$$

The above basic UCB1 formula can be improved [12, 126] in the following way: choose machine j that maximizes

$$\overline{X}_j + \sqrt{\frac{\log n}{T_j(n)} \min\left\{\frac{1}{4}, V_j(T_j(n))\right\}} \tag{3.6}$$

where

$$V_j(s) = \left(\frac{1}{s}\sum_{\gamma=1}^{s} X_{j,\gamma}^2\right) - \overline{X}_{j,s}^2 + \sqrt{\frac{2\log n}{s}} \tag{3.7}$$

is the estimated upper bound of the variance of rewards of machine j.

The extension of UCB1 algorithm applicable to game tree searching is known as UCT [170]. In short, each position is treated as a bandit and each move that can be played in a given position - as an arm. In order to assess the expected payoffs of moves in a game tree, the following scheme is repeatedly applied:

1. start from the root node;
2. until stopping criterion is met do:
   a) choose a move according to UCB1 policy (3.6)-(3.7);
   b) update position (perform a chosen move);
3. score the game (the end-of-game position is reached);
4. update all visited nodes with this score.

The preferable stopping criterion is reaching the end-of-game position, i.e. expanding the paths all the way down to the leaves in the full game tree. The value of each node is computed as the mean of values of all its child nodes weighted by the frequency of visits.

In UCT implementation for Go [125, 126] the tree can be built incrementally in a way proposed by Coulom [72]. In this method the tree is revealed gradually, node by node. Starting from the root node the above described UCT procedure is applied until a previously unseen position is reached. This position (node) is added to the tree and the game is randomly played till the end by "plain" Monte Carlo simulations, then scored and the score is used as the first approximation of the newly added node's value and used to update all its ancestors' estimations. The nodes visited during the Monte Carlo simulation are not added to the tree nor are their values saved. The end-of-game score is not calculated based on the territory score but is either equal to 1 (a win) or 0 (a loss). Draws are ignored since they are very rare in Go.

The UCT implementation is further improved in MoGo by parametrization of the formula (3.6) and by adding the first-play urgency (FPU) adjustment to influence the order of nodes' exploration with respect to rarely visited nodes, e.g. the ones located far from the root node. Moreover, the quality of play is enhanced by parallel implementation of the algorithm and the use of domain knowledge in the form of suitable pattern templates - see [125, 126] for the details.

The effect of applying UCT method to Go is astonishingly good. The strength of MoGo, Fuego, CrazyStone and other Go programs basing on Monte Carlo simulations further confirms the efficacy of roll-out simulations - this time in the case of Go - game with very high branching factor and for which the simple yet reasonable evaluation function has not yet been developed (and is doubtful to exist at all [237]). Mainly due to the above two reasons traditional minimax search methods have only limited applicability to Go. The UCT, on the other hand, can efficiently handle both the large size of the game tree and the lack of compact evaluation function.

The success of the UCT, except for several "technical" improvements, can be attributed to the three following features: first of all, as opposed to alpha-beta search, the algorithm can be stopped at any time and still yield reasonable results. Second of all, the UCT handles uncertainty in a smooth way since the estimate of the parent node is the average of the estimates of its child nodes weighted by their confidences (frequencies of visits). Hence the algorithm prefers the selection of the best child node if the estimate of that node is visibly higher than the estimates of its siblings and the confidence is high. In effect the estimation of the parent node will be highly weighted by the value of this "leading" child node. On the other hand, in the case there is no such favorite node and two or more child nodes have similar estimates, or the confidence is low the resulting estimation of the parent node will be close to the average of the child nodes estimations. The third advantage of the UCT in Go is that the game tree grows incrementally, in an asymmetric manner, with deeper exploration of promising lines.

4

State of the Art

The 1990s brought the renaissance of mind game programming, leading to several extraordinary achievements in machine-human competitions. The best-playing programs surpassed world human champions in chess, checkers, Scrabble, and Othello, and are making a striking progress in Go, poker and bridge.

The volume of research projects and research papers devoted to AI and games is too high to even try to describe the discipline and the achievements in a comprehensive and exhaustive way. Instead, a subjective author's selection of some notable examples of AI achievements in games is proposed in this chapter, keeping the descriptions on a relatively general level without going into details. The idea is to give the reader a flavor of these influential approaches and methods rather than a formal, technical description of them.

Presentation is organized around particular games. Chapter 4.1 is devoted to chess, mentioning the famous Deep Blue machine and the latest accomplishment - Rybka. The next chapter introduces Chinook, the World Man-Machine Champion in checkers. Chapter 4.3 describes Logistello, program which won the Othello World Champion title in 1997. Chapter 4.4 mentions the two most popular Scrabble programs: Maven and Quackle competing nip and tuck with each other and leaving human players far away. The next chapter presents the latest AI achievements in poker and bridge - games in which machines are almost as good as human players, though there is still some edge on humans' side. The final chapter is devoted to Go, one of the latest strongholds of human supremacy over machines left in the mind game area.

4.1 ..., Deep Blue, Fritz, Hydra, Rybka, ...

Since *The Turk* times several serious attempts to build chess playing programs have been noticed, most of them relying on manually defined evaluation functions and high speed full-width alpha-beta search. One of the interesting proposals was Hitech [26], the strongest chess playing program

J. Mańdziuk: Knowledge-Free and Learning-Based Methods, SCI 276, pp. 41–50.
springerlink.com   © Springer-Verlag Berlin Heidelberg 2010

in the late 1980s. Hitech implemented the so-called SUPREM architecture, part of which was a very effective pattern-based search engine. Game patterns together with conditions defining their applicability formulated the game specific knowledge used in the evaluation process.

Presumably the most striking achievement of AI in games was *Deep Blue's* victory over Garry Kasparov - the World Chess Champion (at the time the match was held) and one of the strongest chess players in the history. This event ended a nearly 50-year era of chess programming efforts started by Shannon's paper.

Deep Blue took advantage of an extremely fast search process performed in parallel by 480 specially designed chess chips constituting a 30-node cluster (16 chips per processor). The system allowed total search speed between 100 million and 330 million positions per second depending on their tactical complexity [57] or 50 billion positions in three minutes - the average time allotted for each move [160]. The machine used a very sophisticated evaluation function composed of over 8,000 features implemented in each of the 480 chess chips. On the software side, Deep Blue implemented the Negascout algorithm with transposition tables, iterative deepening, and quiescence search.

Apart from tuning thousands of weights in the evaluation function, a lot of other AI engineering problems concerning massively-parallel, nonuniform, highly-selective search, or creation and analysis of the extended opening book and endgame database, had to be solved in order to achieve the final result. There is no doubt that Deep Blue was a milestone achievement from an engineering and programming point of view [57, 155]. From a CI viewpoint much less can be said, since the system did not take advantage of any learning or self-improvement mechanisms.

The victory of Deep Blue attracted tremendous interest in the game research community and had undisputed social and philosophical impact on other people, not professionally related to science (New York's Kasparov vs. Deep Blue match was followed on the Internet by thousands of people all over the world). On the other hand, the result of the match should not, and apparently did not, hinder further research efforts aiming at developing an *intelligent* chess playing program equipped with cognitive skills similar to those used by human chess grandmasters.

Since Deep Blue's victory several other chess programs, e.g. *Shredder, Fritz, Deep Junior*, or chess supercomputer - *Hydra* played successfully against human grandmasters, but none of these matches gained comparable public attention and esteem. The list of the recent computer achievements includes the success of *Deep Fritz* over the World Chess Champion Vladimir Kramnik and striking ranking results of *Rybka* [105, 261], the state-of-the-art accomplishment in machine chess playing. Rybka - written by Vaclav Rajlich - is a computer program, which outpaces human and computer competitors by a large margin. On official ranking lists Rybka is rated first with extraordinary result of 3238 ELO points (as of September 26, 2008) on Intel Core

2 Quad 6600 with 2GB memory and is approaching 3000 ELO points on a single CPU PC machine[1]!

4.2 Chinook – A Perfectly Playing Checkers Machine

The first computer program that won human world championship in a popular mind game was *Chinook* - the World Man-Machine Champion developed by Jonathan Schaeffer and his collaborators from the University of Alberta [284, 287, 289]. Chinook's opponent in both championship matches in 1992 and 1994 was Dr Marion Tinsley - the ultimate checkers genius, who was leading the scene of checkers for over 40 years. During that period Dr Tinsley lost as few as only 7 games (including the 2 to Chinook)! As Jonathan Schaeffer stated in his book [289], Tinsley was "as close to perfection as was possible in a human."

During the re-match in 1994 Chinook was equipped with a complete 7-piece endgame database (i.e. exact solutions of all endings of 7 pieces or less) and with a 4 × 4 subset of the 8-piece database (i.e. all endings in which each side was left with exactly 4 pieces) [289].

Similarly to Deep Blue, Chinook was a large-scale AI engineering project including all vital aspects of AI design: efficient search, well-tuned evaluation function, opening book and endgame database. Special care was taken of specific tactical combinations (e.g. exchanging one of Chinook's pawns for two of the opponent) which were carefully analyzed and coded in special tables. The evaluation function was of linear form, composed of 25 major components, each having several heuristic parameters. Examples of these component features included piece count, kings count, and runaway checkers (having a free path to promotion). All evaluation function's weights were laboriously hand-tuned.

Four such functions differing by the choice of weights were actually implemented in Chinook. Each set of weights was devoted to particular phase of the game, which was divided into four parts according to the number of pieces remaining on the board. Phase one lasted between 20 and 24 pieces, phase two: between 14 and 19 pieces, phase three: between 10 and 13 pieces, and the last phase commenced when less than 10 pieces were left. Higher weights were, for example, assigned to king's presence in the later game phases than in the earlier ones.

Since the initial developments in 1989, Schaeffer and his collaborators concentrated on building perfect checkers player by solving the game. The result was ultimately achieved almost 20 years later and published in *Science* [286].

---

[1] The ELO rating system is a standard method of calculating the relative strength of a player in two-player board games, including chess and checkers. The highest ever human ranking in chess was 2851 points achieved by Garry Kasparov in July 1999 and January 2000. Only 4 human players in the history ever exceeded the level of 2800 points.

Checkers was proven to be a draw game (assuming perfect play of both sides). As part of the proof a complete 10-piece endgame database was developed. At the moment checkers is the most challenging board game among those solved to date.

4.3   Logistello – The Othello World Champion

Another game in which computers visibly outperformed humans is Othello. In 1997, only three months after Deep Blue's victory, Michael Buro's program *Logistello* [56], running on a single PC machine, decisively defeated the then Othello World Champion Takeshi Murakami with the score 6 − 0. Taking additionally into account that the post-mortem analysis of the games played during the match showed that the program was not in trouble in any one of them, demonstrates Buro's achievement even better.

Unlike Deep Blue and Chinook, which both heavily relied on game-dependent knowledge and game-dedicated methods, Logistello used several game-independent solutions. Three such universal methods that contributed to the program's flawless victory were:

(1) the new, efficient way of feature selection for the evaluation function - GLEM (generalized linear evaluation model) [54]. Starting from a set of predefined, atomic features, various Boolean combinations of these simple features were considered and their weights calculated by the linear regression method based on several million training positions, generated in a self-play mode, labeled either by their true minimax value or an approximation of it. Logistello used 13 such functions devoted to particular phases of the game. The phases were defined based on the number of discs played already in a game.

(2) Forward pruning method ProbCut (and Multi-ProbCut) capable of cutting out the most probably irrelevant subtrees with predefined confidence [53]. In short, the method generalized from shallow search to deeper search levels by statistically estimating the coefficients in the linear model approximating the relationship between shallow and deep minimax search results.

(3) Automatic opening book development, which made use of search results along the promising lines not actually played [55]. The algorithm maintained a game tree, rooted at the initial game position, based on played games, and labeled the leaf nodes according to the previous game outcomes or based on the negamax approximations. Additionally, in each interior node the best continuation among not played moves was found by negamax search and added to the tree together with the corresponding assessment. This tree, composed of both played and promising not-yet-performed moves, was used by Logistello as a basis for automatic development of the new (yet unplayed) openings. This way the program was able to avoid repeating poor paths

played previously, since at each node there existed a plausible alternative to be considered.

All three above-mentioned aspects of game playing are applicable also to other two-player board games. On a more general note these methods are in line with human way of playing which includes autonomous building of the evaluation function, performing selective search, and looking for new variants in the known game openings. In the spirit of this book, Logistello is actually regarded as one of the chronologically first examples of successful application of a more human-like approach (consisting in autonomous learning and generalization) to game playing, in a master-level program.

4.4 Maven and Quackle – The Top Level Machine Scrabble Players

One of the first programs that achieved a world-class human level in a non-trivial game was *Maven* - a Scrabble playing program written by Brian Sheppard [298].

Scrabble is a game of imperfect-information with a large branching factor (the average number of possible words that can be played in a given position is around 700), and as such is very demanding for AI research. The goal of the game is to form words on a 15×15 board using the rules and constraints similar to those of crossword puzzles. Each of the players maintains a set of his private tiles, which he uses to make moves (form words on the game board). The exact rules of the game and the scoring system can easily be found on various Internet sites.

In the 1990s Maven successfully challenged several world top Scrabble players including (then) North America Champion Adam Logan and world champion Joel Sherman (both matches took place in 1998). Strictly speaking the supremacy of Maven was demonstrated only in North America - i.e. for US English, but adaptation of Maven to another dictionary is straightforward. Actually, it was later adapted to UK English, International English, French, Dutch, and German.

The program divides the entire game into three phases: a normal game, a pre-endgame and an endgame. The pre-endgame starts when there are 16 unseen tiles left in the game (7 on the opponent's rack and 9 in the bag)[2]. The endgame commences with drawing the last tile. The key to Maven's success lies in efficient selective move generation and perfect endgame play supported by B^* search algorithm[3] [25]. Maven uses game scenarios simula-

---

[2] Maven as well as other world-class programs is developed for a two-player variant of the game. In a general case Scrabble can be played by 2-4 players.

[3] As soon as the bag is empty, Scrabble becomes a perfect-information game, since one can deduce the opponent's rack by subtracting the tiles one can see from the initial distribution.

tion or "rollouts," discussed in chapter 3.3, which proved to be very effective in this game. The implementation details are presented in [298].

In [298] Sheppard stated: "There is no doubt in my mind that Maven is superhuman in every language. No human can compete with this level of consistency. Anyone who does not agree should contact me directly to arrange a challenge match." So far, no-one tried.

Recently, another very strong computer Scrabble player has entered the scene. The crossword puzzle solving software named Quackle[4] [166] outplayed Maven during the 2006 Human vs. Computer Showdown in Toronto. Both machines played against the same 36 human opponents achieving excellent scores: 32-4 and 30-6, respectively. Outplaying Maven gave Quackle the right to play a best-of-five match against the best human contestant David Boys. Boys - the former Scrabble World Champion (in 1995) - was outwitted by Quackle losing three out of five games. The Quackle software is released under the GPL, i.e. the source code is freely available. This makes a great opportunity for all interested in the development of a world class Scrabble AI player.

Quackle exploits similar ideas as Maven. The choice of a move is decided based on some number of simulations, each of them ending with assessment of the resulting board made with the use of a static evaluation function. Simulations are restricted to a small subset of the most promising moves in a given position. During the simulations each opponent's move (word to be played) is composed of the tiles from the opponent's rack. The rack is randomly sampled as a 7-tile set out of the remaining tiles. A set of remaining tiles is calculated as the difference between the initial set of tiles and the tiles already played in the game and the tiles possessed by the player.

The results of Quackle and Maven against top human players constantly prove that the gap between computers and humans is unlikely to be narrowed down. Computers are capable of memorizing huge dictionaries, which together with efficient search and evaluation algorithms make them exceptionally good players. In this context the competition between programs and people seems to be less and less challenging for machines.

Despite achieving a world-class Scrabble competency the question that can be posed is whether machines can further improve their playing skills and eventually become infallible, perfect players. Paradoxically, one of the possible avenues to explore is becoming a more human-like player in some aspects. In particular, programs may potentially follow humans in implementing the process of modeling the opponent's rack and consequently considering not only the moves which lead to maximal profit, but also the ones that prevent the opponent from scoring high, by blocking or destroying his opportunities. This direction has been recently explored by Mark Richards and Eyal Amir, the authors of a "dirty playing" Scrabble-bot [266]. The key feature of the system is smart prediction of the opponent's rack, which allows prediction

---

[4] Due to some legal issues Quackle is released as a crossword puzzle program, which has officially nothing in common with Scrabble.

of opponent's possible moves, the best of which can in turn be possibly fore-
stalled by bot's blocking moves. The above concept is further discussed in
chapter 11 entirely devoted to the problem of modeling the opponent and
handling the uncertainty.

4.5 Poker and Bridge – Slightly behind Humans

Poker

Poker is an imperfect-information card game. The existence of hidden information and additional difficulties attributed to the frequent use of various kinds of deception (bluffing), when the game is played by experienced
humans, make efficient poker playing a really challenging task for artificial
systems. Inaccessibility to the whole information requires effective modeling
of the opponent's decision making process, applying risk management techniques and using dynamical, context sensitive and, in most cases, probabilistic
evaluation functions rather than static ones. One of the pioneering top-class
poker programs is *Poki-X* [31, 280] written by Jonathan Schaeffer and his
research group from the University of Alberta. Poki uses a sophisticated,
multi-stage betting strategy which includes the evaluation of *effective* and
potential hand strengths, opponent modeling and probabilistic simulations
based on Monte Carlo sampling. All the above techniques are essential for
successful machine poker playing. For the scope of this book the problem of
opponent modeling (discussed further in chapter 11) is of particular interest.

The latest achievement (though probably not the last word) of the University of Alberta poker group is a Texas Hold'em system - *Polaris*[5]. In July
2007 Polaris played against two top poker professionals in the so-called duplicate match (the same cards are played twice in a human-bot pair, but with
the reversed seating). The competition consisted of 2,000 duplicate hands
divided into 4 separate matches. After 16 hours of play Polaris appeared to
be only slightly inferior to people by winning one, drawing one, and losing
two rounds. One year later Polaris took part in the Second Man-Machine
Poker Championship, where it challenged six human professionals winning
three, drawing one, and losing two out of the six matches.

The results of Polaris (and also other, less acquainted poker bots, e.g. [136])
indicate that machines are practically playing on par with the best human
players in the two-player so-called Limited Hold'em variant of the game,
where amount of money put at stake at each betting decision is limited. In the
No Limit version there are still some lessons to be learned by computer players
in order to catch up with people. One of the main reasons why people are still
ahead of the machines in the case of No Limit poker is the humans' perfect
ability to bluff at the level which is still unavailable to machines. Another
prospective challenge for current poker bots is the multi-player version of the

---

[5] For the references to Texas Hold'em see the footnote in chapter 3.3.

game, which involves several *strategic* and *psychological* issues not existing in the much simpler, two-player case.

Bridge

The other card game very popular among humans and also a longstanding target for AI/CI game research is bridge. Bridge is imperfect-information game, played in pairs. Players in a pair cooperate with each other and the performance is measured as the playing quality of the whole pair. In tournament situations, in order to eliminate the "luck factor" each deal is played twice with swapping sides between pairs. Hence, a bridge team is typically composed of four players divided into two pairs. The rules of bridge can be found on several Internet game sites. A summary of the rules is also presented in [135]. A concise review of previous AI and CI attempts in computer bridge can be found in [235].

Computers officially compete in bridge contests since 1997 when the annual World Computer-Bridge Championship has been organized for the first time. The regular participants in this event are *Jack*, *Bridge Baron*, *WBridge5*, *Micro Bridge*, *Q-Plus Bridge*, and *Blue Chip Bridge*. Each of these programs enjoyed some success in previous contests, but the most renowned one is the French program WBridge5 [343], the winner of World Computer-Bridge Championships in 2005, 2007 and 2008. A Dutch program named Jack, written by Hans Kuijf and his team [176] is a close runner-up. Jack won the title four times in a row in 2001 − 2004 and for the fifth time in 2006. Both WBridge5 and Jack are commercial programs and very little is publicly known about their internal construction and the algorithms they use. The level of play of both programs is gradually improving and currently they are able to play on almost equal terms against top human bridge players.

An interesting phenomenon among top bridge programs was GIB (Ginsberg's Intelligent Bridge Player) written by Matthew L. Ginsberg - historically the first strong bridge playing program and the winner of the 1998 and 1999 World Computer Bridge Championships. The underlying idea of GIB was to take advantage of the ability to analyze and solve the DDBP - a perfect-information variant of bridge, introduced in chapter 3.3. In order to do that GIB used partition search, a cutting tree technique defined by Ginsberg, which generalized the idea of transposition tables to the case of *sets* of game situations instead of individual cases. Similar sets of game situations (full 13-card hands or a few card game endings) were stored together with associated common estimations. Observe that the idea of partition search is particularly well suited to the game of bridge. In many game situations the exact placement of lower ranked cards is not important and what matters is only the *number* of such cards on each hand, not their particular assignment. Situations of this type can be naturally grouped into sets of common, in some sense indistinguishable, game states.

Having the ability to efficiently solve the DDBP, GIB used Monte Carlo-based selective sampling to generate various cards' distributions and played them virtually, using the partition search method, as discussed in chapter 3.3. Based on simulation results the actual game decision (a betting bid or selection of a card to be played) was made.

The level of play of top bridge programs has been steadily improving over the last $10 - 15$ years and has recently reached that of expert human players. Although there is still some work that needs to be done, especially in improving the bidding abilities, in order to catch up with human champions, the gap is quickly narrowing down and human supremacy in this game may be seriously challenged within a few years.

4.6 Go – The Grand Challenge

Go is undoubtedly one of the grand AI/CI challenges in mind game domain[6]. Despite simple rules and no piece differential, playing the game at the level comparable to that of human masters is yet a non-achievable task for machines. There are several reasons for this situation. First of all, Go has a very high branching factor, which effectively eliminates brute-force-type exhaustive search methods. A huge search space, however, is not the only impediment in efficient computer play.

The very distinctive feature that separates Go and other popular board games is the fact that static positional analysis of the board is orders of magnitude slower in Go than in other games [237]. Additionally, proper positional board judgement requires performing several auxiliary searches oriented on particular tactical issues. Due to the variety of positional features and tactical threats it is highly probable that, as stated in [237], "no simple yet reasonable evaluation function will ever be found for Go."

Another demanding problem for machine play is the "pattern nature" of Go. On the contrary to humans, who possess strong pattern analysis abilities, machine players are very inefficient in this task, mainly due to the lack of mechanisms (either predefined or autonomously developed) allowing flexible subtask separation. The solutions for these subtasks need then to be aggregated - considering complex mutual relations - at a higher level and provide the ultimate estimation of the board position. Instead, only relatively simple pattern matching techniques are implemented in the current programs [236, 237].

Due to still not very advanced stage of Go playing programs' development it is hard to point out the stable leader among them. Instead, there exists a group of about ten programs playing on a more or less comparable level. These include: *Many Faces of Go, MoGo, Fuego, Zen, CrazyStone, Golois, GoIntellect, Explorer, Indigo,* and a few more.

Recently, a lot of interest and hope have been devoted to the Monte Carlo selective sampling methods, which are perfectly suited to games with high

---

[6] Unless stated otherwise we are referring to the game played on a regular 19 x 19 board.

branching factors. The idea of applying MC sampling-based UCT algorithm (presented in section 3.3.1) was implemented in MoGo [125, 126, 340] and Fuego [92], establishing the programs as current leaders among computer players.

A detailed discussion on the development of Go playing agents can be found in [40, 237]. The researchers aiming to write their own Go programs may be interested in analyzing the open source application Gnu Go [48].

Part II

CI Methods in Mind Games. Towards Human-Like Playing

An Overview of Computational Intelligence Methods

Computational Intelligence is a relatively new branch of science which, to some extent, can be regarded as a successor of "traditional" Artificial Intelligence. Unlike AI, which relies on symbolic learning and heuristic approaches to problem solving, CI mainly involves systems that are inspired from nature, such as (artificial) neural networks, evolutionary computation, fuzzy systems, chaos theory, probabilistic methods, swarm intelligence, ant systems, and artificial immune systems. In a wider perspective CI includes also part of machine learning, in particular the reinforcement learning methods.

Applying either one or a combination of the above-mentioned disciplines allows implementation of the elements of learning and adaptation in the proposed solutions which make such systems somehow intelligent.

In the game playing domain the most popular CI disciplines are neural networks, evolutionary and neuro-evolutionary methods, and reinforcement learning. These domains are briefly introduced in the remainder of this chapter along with sample game-related applications. The focus of the presentation is on the aspects of learning and autonomous development. Relevant literature is provided for possible further reading.

5.1 Artificial Neural Networks

Artificial neural networks or *neural networks* (NNs) in short, are computational structures capable of processing information (provided as their input) in order to accomplish a given task. A NN is composed of many simple elements (called neurons) each of which receives input from selected other neurons, performs basic operations on this input information and sends its response out to other neurons in the network.

An inspiration for the above way of processing information is biological nervous system and in particular biological brain. NN models can therefore be regarded as very crude simplification and abstraction of biological networks.

J. Mańdziuk: Knowledge-Free and Learning-Based Methods, SCI 276, pp. 53–70.
springerlink.com   © Springer-Verlag Berlin Heidelberg 2010

There are several types of NN models. Depending on particular task to be solved an appropriate neural architecture is selected. The basic division of neural models can be made according to the training mode, i.e. supervised vs. unsupervised. A typical example of the supervised network is Multilayer Perceptron (MLP) described below in section 5.1.1. An example of unsupervised network is Self Organizing Map [171]. Another possibility is to divide neural architectures based on the information flow direction into feed-forward (e.g. MLP), partly recurrent (e.g. Elman [89] or Jordan [162] networks), and recurrent (e.g. Hopfield models [151]).

Neural networks have been successfully applied to various recognition and classification tasks [33], optimization and control problems [152, 205], medicine [202], financial modeling and prediction [225], bioinformatics and chemistry [192], games [13], and many other fields. Due to variety of possible applications NNs are often regarded as *the second best way of solving any control/classification problem.* In other words, for any given problem at hand a dedicated heuristic approach will most probably be more effective than a NN-based solution, but on the other hand, the generality of NN methodology allows its application to a wide spectrum of problems.

The reader interested in other than games applications of NNs is advised to refer to the above cited books and articles. Fundamentals of neural networks can be studied for example by reading the books by Bishop [33], Hassoun [146], or Haykin [147].

5.1.1 Multilayer Perceptron

Multilayer Perceptron is the most popular type of feed-forward neural networks, which in turn are the most popular neural architectures (in general and in particular in game domain). Neurons in an MLP follow McCulloch-Pitts

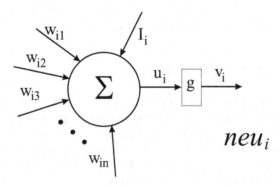

Fig. 5.1. A model of McCulloch-Pitts neuron. See description within the text

neuronal model [224] presented in fig. 5.1. The dynamics of the ith neuron (denoted by neu_i in the figure) is defined by the following set of equations:

$$\begin{cases} u_i(t) = \sum_{j=1}^{n} w_{ij} v_j(t-1) + I_i \\ v_i(t) = g(u_i(t)) \end{cases} \tag{5.1}$$

where u_i, v_i denote input and output signals of the ith neuron, respectively, I_i is constant external bias signal, w_{ij} is the weight of connection from the jth neuron to the ith one, g is nonlinear transfer function and t denotes time step. In the case of binary neurons g is usually a step function and in the continuous case g is sigmoid-shaped, i.e. respectively

$$g(u) = \begin{cases} 1 & \text{if } u \geq 0 \\ 0 & \text{if } u < 0 \end{cases} \tag{5.2}$$

or

$$g(u) = \frac{1}{1 + e^{-\alpha u}} \tag{5.3}$$

for some positive coefficient α.

Neurons in the MLP (fig. 5.2) are structured in layers: the input layer (which receives the input signal), the so-called hidden layers (in majority of applications there are between zero and two such layers) and the output layer (which yields the result). In the MLP a signal is forwarded in one direction only - from the input layer though the hidden layers to the output layer. The signal is propagated in a synchronous manner, one layer at each time step. Connections are allowed only between neighboring layers. There are no loops or feedback connections. Each connection has a weight assigned to it. It can be either positive (excitatory connection), negative (inhibitory one) or equal to zero (no connection). Following the McCulloch and Pitts neuron's scheme the input signal received by each neuron in a given layer is a weighted sum of output signals - multiplied by connection weights - from all neurons belonging to the preceding layer. This weighted input is then transformed with the use of nonlinear function g of the form (5.2) or (5.3) and propagated up to the next layer's neurons. The input layer only forwards the received input signal without any mathematical or functional modification.

How can one use an MLP-based computational model to learn a given task (being either classification, approximation, or control problem)? Not going into details, let's treat the network as a black box for a moment and suppose that the goal is to approximate a given unknown mapping $f : \mathcal{R}^n \rightarrow \mathcal{R}^m$ having sufficient number of examples of the form $(X^i, Y^i) = (x_1^i, x_2^i, \ldots, x_n^i; y_1^i, y_2^i, \ldots, y_m^i)$, $i = 1, \ldots, p$, where $(X^i, Y^i) \in \mathcal{R}^n \times \mathcal{R}^m$ are pairs of input and output vectors.

The MLP can be trained by a backpropagation (BP) method [273, 274, 344] or any of its newer variants, e.g. RProp [268] or QuickProp [100]. In short, the BP method relies on iterative improvement of the connection weights so as to minimize the error function calculated over the training set (X^i, Y^i),

Multilayer Perceptron

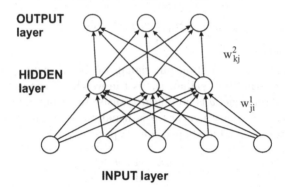

Fig. 5.2. A Multilayer Perceptron with one hidden layer. Weight w_{ji}^r, $i = 1,\dots,n_{r-1}$; $j = 1,\dots,n_r$, $r = 1,2$ is assigned to connection from neuron neu_i in the $(r-1)$th layer to neuron neu_j in the rth layer, with the input layer denoted as the zeroth layer. n_0, n_1 and n_2 are the numbers of units (neurons) in the input, hidden and output layer, respectively

$i = 1,\dots,p$. This error function is typically defined as mean square error of the NN output:

$$E(w) = \frac{1}{2}\sum_{i=1}^{p}\sum_{k=1}^{m}(v_k^i - y_k^i)^2, \tag{5.4}$$

where w denotes the connection weights matrix, p - the number of training patterns, $v^i = [v_1^i,\dots,v_m^i]$ and $y^i = [y_1^i,\dots,y_m^i]$ are, respectively, actual and desired neural network output vectors related to the training pair (X^i,Y^i), $i = 1,\dots,p$. Connection weights modifications rely on calculation of the gradients of the error function with respect to these weights.

With differentiable activation functions employed in all neurons, the required gradients may easily be obtained via a backpropagation procedure, which involves reversal of the network's signal flow direction, replacement of activation functions with linear transformations using their derivatives as slope coefficients and application of difference between actual and training output vectors as input to this reversed network. Required derivatives of the error function with respect to individual connection weights can then be computed by multiplying activation values of neurons acting as input elements of each of these connections in original and reversed networks.

In its simplest form, BP method uses simple gradient descent algorithm to calculate connection weight changes. Weights are therefore modified according to the following formula:

$$\Delta w(t) = -\eta \nabla E(w(t)), \tag{5.5}$$

where η is a learning rate coefficient. Some improvement can be achieved by adding a *momentum* term $\mu \in [0, 1]$:

$$\Delta w(t) = -\eta \nabla E(w(t)) + \mu \Delta w(t - 1), \tag{5.6}$$

which in the case of low value of $\nabla E(w(t))$ (which happens for example when the process enters a plateau of E or in the neighborhood of a local minimum) speeds up the learning process and prevents possible oscillations. Other, more sophisticated gradient algorithms, such as RProp [268] and QuickProp [100], are, however, often used to good effect.

Potential strength of NNs application to the problems requiring approximation of unknown function has been proven theoretically and MLPs are known to be *the universal approximators*. It can be proven that a one-hidden-layer feed-forward neural network with sigmoid neurons is capable of approximating any continuous, multivariate function to any desired degree of accuracy, and that failure to map a function arises from poor choices of network's parameters (weights, biases, or slope α in (5.3)) or insufficient number of hidden neurons [73, 153, 154]. In other words, given a sufficient number of hidden sigmoid units the error of approximation of a given continuous nonlinear function by the one-hidden-layer MLP network can be arbitrarily low.

Approximation capabilities of feed-forward neural networks provide a solid theoretical basis for application of these architectures to the tasks consisting in efficient function approximation. Unfortunately the universal approximation theorem is only existential, i.e. it does not provide a recipe for construction of the optimal architecture for a problem at hand. The choice of the MLP's parameters and the number and cardinalities of hidden layers are in most tasks a matter of an educated guess or a trial-and-error approach, based on the experience of the designer.

5.1.2   Neural Networks and Games

In game domain NNs are a popular way to represent an evaluation function. A neural architecture used in this task is typically a feed-forward network implemented as an MLP or its modification. The topology of the network is usually tuned to the specific game. In many cases some of the weights in the network may be fixed (i.e. non-trainable) - often to make some game-specific concepts, such as particular spatial arrangements, obvious to the network. The network's input layer is often extended to include not only the raw representation of a game state, but also a number of predefined game features.

The MLP is usually trained with a variant of backpropagation method. Typically, the learning process is based on successive presentation of the training vectors according to some fixed, randomly chosen order, in an iterative multiple-epoch manner. For a given neural architecture and available set of training vectors, the result of the process depends mainly on the three following factors: the choice of internal training parameters - especially the learning rate and the momentum, the choice of initial networks' weights, and the order of presentation of the input vectors within each epoch. Quite a lot of attention was devoted to the problem of optimal choice of the internal training parameters in various applications other than games. In game-related papers these parameters were usually chosen by an educated guess or based on some non-exhaustive trial and error procedure, which generally proved to be sufficiently effective.

The initial weights are most often chosen as small random values from an interval $[-a, +a]$, for some positive a, with the concern that too small value of a may hinder the learning process (due to small weight updates), whereas choosing a too large may prevent the learning process convergence (due to explosive growth of the modified weights). In practice, if the results of the learning process are unsatisfactory, it may be repeated with another choice of initial weights and/or algorithm's parameters.

The last issue relates to the order of presentation of the training data. Unless additional information about the distribution of the input vectors is available, the order of patterns within the first training epoch is usually chosen at random, and remains fixed for all subsequent epochs. Some research results concerning more efficient ordering of the input patterns were published in domains other than games, but to the author's knowledge this topic is almost unsurveyed in game area.

Recently, the idea of *interleaved training* (described in detail in chapter 10.4), in which a random sequence of input patterns is interleaved with a specially ordered sequence (of these same patterns), was proposed by the author [82] in chess domain. The work is yet at the preliminary stage, but hopefully will lead to some interesting observations.

5.2   Evolutionary Methods

Evolutionary methods [110, 140, 150] compose a special class of optimization algorithms. At their core, their task is finding a minimum or maximum value of a function over a specified solution space. They include a class of procedures which attempt to mimic adaptation processes observable in the nature, as all living organisms struggle to perform in the environment better than both their peers and other species. Genetic algorithms [140] is one group of such procedures, distinguished by its popularity and simplicity.

Evolutionary methods in general belong to population-based algorithms. They start with generation of a set of solution candidates. Often, this is simply a group of totally random entities - usually referred to as *individuals* or *specimens*, forming a *population*. Each candidate solution is typically described by a vector of parameters, called a *genotype*. The initial population is afterwards transformed in an iterative process, referred to as *evolution*, until some *a priori* defined stop condition is fulfilled. This process typically involves – in some form – some or all of the biological evolution inspired operations of selection, crossover, and mutation.

The aim of *selection* process is identification of a subset of the population containing promising individuals that should be allowed to breed a new generation of candidate specimens. This process is driven by the concept of individuals' *fitness comparison*. In most implementations, fitness of an individual is represented as a numerical value assigned to it by a problem-specific *fitness evaluation function*. In typical evolutionary methods selection is a stochastic process in which fitter candidate solutions are more likely to be chosen.

Crossover (or *recombination*) and *mutation* procedures form the reproduction phase of evolutionary algorithm. New offspring individuals are created during this phase, based on parent specimens chosen during the selection phase. A crossover operator defines a way to combine description of a pair (or sometimes more than two) of parent specimens in order to create one or several offspring candidate solutions demonstrating a combination of traits of both parents. A mutation procedure amplifies the population diversity by introducing additional random changes into the individual characteristics.

Depending on the evolution strategy implemented, all thus created individuals may be included in the next generation population or some additional selection procedure may be invoked to decide which of the individuals are prospective enough to be left in the pool. Previous population individuals may also either be disposed or maintained in the population, depending on sophisticated filtering methods.

5.2.1 Evolution Strategies

In *evolution strategies* (ES) [27, 28] (used in many evolutionary programming applications) selection of individuals is based not on their actual fitness values, but on their fitness ranking only. In the simplest case (also known as pseudo-evolutionary or local hill-climbing method [174]) a population of two individuals is considered. The current solution (a parent) competes with its mutated version (a child). A child becomes the actual solution only if better (sometimes: not worse) fitted than the parent individual. Otherwise the candidate solution (a child) is rejected. Such a procedure is denoted as $(1+1)$-ES (one parent competes with one mutated version of itself) and is a specific case of a more general $(\mu + \lambda)$-ES, where λ offspring (mutated versions of μ parents) are generated and compete with these μ parents to form the next generation consisting of μ individuals. The $(\mu + \lambda)$-ES is an *elitist* strategy, i.e. the best fitted individual (whether a parent or a child) is always promoted as the parent for the next generation.

A different approach is taken in *non-elitist* strategies, denoted by (μ, λ)-ES, in which λ offspring (mutated versions of μ parents, $\mu < \lambda$) are generated and compete with one another for μ slots in the next generation, whereas the current parents are all disregarded. All of the above are particular realizations of two canonical forms of ES, namely the $(\mu/\rho, \lambda)$-ES and $(\mu/\rho + \lambda)$-ES, respectively. Parameter $\rho \leq \mu$ denotes the mixing number, i.e. the number of parents involved in recombination operation. In the case of $\rho = 1$ no recombination takes place in the evolutionary process and the equivalent shorthand notation (μ, λ)-ES and $(\mu + \lambda)$-ES, respectively, is used.

Contemporary derivatives of the above strategies often use a combination of the above *comma* and *plus* selections denoted by $(\mu/\rho_,^+\lambda)$-ES.

5.2.2   Evolutionary Methods and Games

Evolutionary algorithms are widely used for development of game playing agents. These agents usually occur in one of two forms: either as state evaluators or as action selectors [197]. The former is more popular and represents simply a static game state evaluation function. The latter's task is choosing a move to make upon being presented the current game state. Implementation of evolutionary algorithm for action selector agent is in most cases much more difficult.

Several possible forms of game agents' representation have been employed in various evolutionary experiments. In the simplest of them, each solution would be represented as a set of coefficients for computing evaluation values as linear combination of predefined game state numerical features – such as the number of stones or mobility of players. More sophisticated approaches would incorporate more sophisticated rules. Yet, lately MLP-based solutions have become a popular form of game playing agent representation. The MLPs with predefined spatially-aware architectures have been successfully evolved for chess [114], checkers [109] and Othello [70]. Some attempts have also been made to evolve not only the MLP connection weights, but also the topology of the network [230, 231]. The above accomplishments are presented in the next chapter.

Fitness Assessment Strategies

Fitness assessment of an individual can be based either on computing assessment error across a set of previously analyzed game positions (e.g. extracted from expert games) or upon results of games played by it against a number of opponents. The former requires a significant amount of expert knowledge when defining suitable evaluation function, therefore the latter choice, although computationally-intensive, is very popular. In some cases external set of opponents may be used for the tournament fitness assessment, but it is common to hold a tournament between agents within the evolved population (or agents from several simultaneously evolved populations).

The latter case is known as coevolution and offers the flexibility introduced by not requiring any externally provided playing agents. Additionally, coevolution avoids problems caused by the need to tune opponent quality to the current capabilities of evolving individuals. On the other hand, in typical implementations of coevolution, fitness assessment function effectively changes from generation to generation, introducing the moving target problem. A coevolutionary learning process lacks also additional guidance provided by elements of external knowledge.

One of possible other approaches to fitness assessment of individuals representing heuristics for checkers and give-away checkers, without external expert knowledge, was presented in [177, 178] and [208]. The authors propose a training scheme in which it is possible to explicitly define fitness function according to which individuals can be compared. In order to achieve that goal, the learning process must be divided into several phases, each corresponding to a separate game stage distinguished by the number of moves already performed.

In the first phase, an evolutionary procedure is to generate an evaluation function capable of efficiently assigning minimax values to end-of-game positions. The procedure starts with generation of a number of random game states close to the end of game (i.e. reached after performing a sufficient number of moves). Most such positions can be fully expanded (until the end-of-game states) by one of the minimax search algorithms (typically alpha-beta pruning), without using any heuristic evaluation function. Those that cannot be, are considered to lead to a draw. That way a set of positions with minimax values assigned to them is obtained. Individual's fitness value can therefore be easily defined as inversion of some error measure of evaluation of the positions in this training set. Once this computationally-efficient fitness function is defined, any evolutionary method may be used to obtain a heuristic evaluation function, which, if trained over a suitably large set of examples, should be able to approximate minimax values of all game states in this stage of the game.

Once the first phase is over, a new training set is generated. It should contain both game positions from the game stage already analyzed and game state representations from the preceding game stage - i.e. closer to the game tree root. These positions could then be evaluated by an alpha-beta algorithm with depth limit sufficient to always reach the subsequent game stage, so that heuristic evaluation function obtained in the preceding training phase could be applied. Once the training set is generated and evaluated, an evolutionary algorithm may be employed to approximate evaluation function for game states from all game stages analyzed so far. This procedure may be repeated for all game stages until game tree root is reached and final evaluation function is generated.

The evolutionary procedure described above was applied to checkers and give-away checkers, using simple linear and nonlinear evaluation function representations. Comparisons with traditional coevolutionary approach indicated superiority of this new multi-stage method. Being much less computationally-intensive, it was able to operate on much larger populations and lead to results of higher quality.

Evolution Algorithms' Enhancements

In the course of research in the area of evolutionary computation in game domain, several evolution algorithms' enhancements have become part of the standard toolbox. One of such standard enhancements is *speciation*. Its aim is dealing with the problem of premature convergence of candidate solutions population to a local extremum of a fitness function. When no speciation mechanism is introduced, evolutionary process may cause all individuals to become very similar and cluster in one promising area of solution space without proper exploration, thus hindering identification of the global optimum. This problem may be further aggravated in case of coevolution when the fitness function itself changes with time and premature convergence may be a very serious threat. Speciation's aim is an introduction of additional evolutionary pressure encouraging differentiation of individuals.

One of typical implementation schemes for speciation is *fitness sharing*. Fitness sharing involves introduction of a metric to solution space and computation of similarity of individuals in the evolved population. Individuals surrounded by big numbers of similar candidate solutions are afterwards punished by having their fitness value decreased proportionally to the similarity to their peers. This modification does not prohibit the evolution process from exploiting any identified promising solution space area, while at the same time promoting better exploration of the rest of the problem space. Darwen, Yao and Lin [75, 351] researched the concepts further, suggesting that properly designed speciation may additionally be used as an automatic modularization procedure. They argue that populations generated by evolutionary methods may actually contain much more useful information than is available in the single best individual usually extracted from them as the process result.

Rosin and Belew [272] introduced another enhancement especially useful in case of coevolution, called *hall of fame*. In typical implementations of evolutionary methods, in order to survive an individual has to perform well in each and every generation. Weak solutions are quickly eradicated and their genotype disappears from the population. This may, however, mean that some useful knowledge contained therein gets lost. In case of coevolution, it may happen that the individual performing poorly when confronted with other individuals in a specific generation, might still have an edge over population from another point in time. Therefore, it would be advisable to avoid losing its knowledge. Rosin and Belew propose introduction of additional store of individuals, as a partial remedy to the problem. Their algorithm saves in each generation the best performing individual in a so-called hall of fame. Sample individuals from the hall of fame are afterwards used as additional opponents in the process of fitness evaluation.

Another enhancement proven to significantly influence the quality of coevolution of neural networks in game domain is *parent-child averaging* [258, 275]. It turns out that in many cases it is beneficial to modify the evolution strategy, so that each next generation contains individuals that are weighted averages of parents and their children. The process of parent-child averaging may be represented by an assignment $\overrightarrow{w'} := \overrightarrow{w} + \beta(\overrightarrow{w'} - \overrightarrow{w})$, where $\overrightarrow{w}$ denotes the weights of the parent net, $\overrightarrow{w'}$ – the weights of its best child, and β is a constant factor controlling the averaging process. The idea of parent-child averaging is further discussed in chapter 9.5.

5.3 Reinforcement Learning

Reinforcement learning (RL) is a concept of programming via means of reward and punishment signals generated by the environment in response to the agents' actions. RL makes use of no training patterns required in supervised learning methods - the agent is expected to learn appropriate behavior via trial and error, with no information on how the task should actually be performed.

In the standard model, at each step of the interaction the agent receives an input signal i from the environment indicating in some way the current state of the environment. Basing on this input data, the agent selects an action a to perform. It communicates its decision to the environment, thus changing its state. A scalar value of this transition is then communicated to the agent as a reinforcement signal r. Long-term maximization of this reinforcement constitutes the goal of the agent. The environment may, in general, be nondeterministic, i.e. the same actions performed in the same states may lead to different reinforcements and resulting environment states. It is, however, typically assumed that the probabilities of individual possible state transitions remain constant, i.e. the environment is *stationary*.

One of the important parts of reinforcement learning problem specification is definition of the optimal behavior, i.e. formulation of the function whose value is to be maximized by the agent. The typical approaches include a *finite-horizon model*, an *infinite-horizon discounted model*, and an *average-reward model* [163]. In the finite-horizon model the agent is expected to maximize its expected reward over a fixed number of h consecutive steps:

$$E\left(\sum_{t=0}^{h} r_t\right). \tag{5.7}$$

The value of h may either remain constant or be reduced after each step. This strategy is useful when the duration of the experiment is known *a priori*.

The infinite-horizon model expects the experiment to last infinitely long and takes all the received rewards into account, discounting them, however, according to the discount factor γ ($0 \leq \gamma \leq 1$):

$$E\left(\sum_{t=0}^{\infty} \gamma^t r_t\right). \tag{5.8}$$

Lastly, the average-reward model expects the agent to maximize its long-term average reinforcement value:

$$\lim_{h \to \infty} E \left(\frac{1}{h} \sum_{t=0}^{h} r_t \right). \tag{5.9}$$

One of the interesting aspects of reinforcement learning is that it does not require strict separation of the learning and exploitation phases. As the agent does not rely on any predefined training patterns, but only problem-inherent environment reinforcement signals, it may improve its behavior rules indefinitely and even adapt to a (slowly) changing environment. This, however, brings about the exploration-vs.-exploitation dilemma, as a deterministic agent following a fixed set of rules will not be able to compare its effectiveness with alternative behavior patterns. This issue is significant enough to deserve separate discussion and will be revisited in section 9.1.4.

5.3.1 Temporal Difference Algorithm

Temporal difference (TD) is one of the most popular and successful reinforcement learning algorithms. Although its earliest widely-known application was Samuel's checkers program [277], it was first formally described by Sutton [313]. Sutton, in his seminal article, considered a class of multi-step prediction problems in which the correctness of the prediction is fully revealed only after several steps of prediction. A typical example of such a problem is a weather forecasting scenario, in which the agent is expected to predict Saturday's weather while first being presented Monday's weather data, then Tuesday's, Wednesday's and so on. This problem can be solved using traditional supervised-learning approach in which training set consists of tuples created by pairing each day's measurement data with actual Saturday's weather. This approach ignores, however, the sequential nature of the data and the correlation between consecutive days' measurement data. Sutton argued that the TD method might prove to be more beneficial in this case.

In one of its forms, TD algorithm may be just a reformulation of the traditional supervised-learning algorithm, with gradient-based update procedure:

$$\Delta w = \sum_{t=1}^{m} \Delta w_t, \qquad \Delta w_t = \alpha(z - P_t)\nabla_w P_t, \tag{5.10}$$

where α is a learning rate parameter, z - the proper outcome value, P_t - value predicted in step t, and $\nabla_w P_t$ is a vector of partial derivatives of P_t with respect to w. Noticing that error value $(z - P_t)$ can be calculated as $\sum_{k=t}^{m}(P_{k+1} - P_k)$ (where $P_{m+1} := z$) allows the learning rule to be reformulated as:

$$\Delta w = \sum_{t=1}^{m} \Delta w_t, \qquad \Delta w_t = \alpha(P_{t+1} - P_t)\sum_{k=1}^{t} \nabla_w P_k. \tag{5.11}$$

This weight modification rule represents the TD(1) algorithm. Although in case of linear prediction functions its effects are equivalent to using supervised learning, it has the additional advantage of allowing an incremental computation, as the value of z is only required for computing value of Δw_m. Previous Δw_t computations require only knowledge of two subsequent prediction values and the sum of previous gradient vectors (it is not necessary to store each vector separately).

TD(1) method can be further generalized to a whole class of TD(λ) methods, making use of additional decay parameter $\lambda \in [0, 1]$:

$$\Delta w = \sum_{t=1}^{m} \Delta w_t, \qquad \Delta w_t = \alpha (P_{t+1} - P_t) \sum_{k=1}^{t} \lambda^{t-k} \nabla_w P_k. \tag{5.12}$$

The distinctive feature of all the TD methods is their sensitivity to the difference between subsequent predictions rather than between each prediction and the final outcome. While the TD(1) procedure (i.e. TD(λ) with $\lambda = 1$) assigns the same weight to all predictions, all temporal difference schemes with $\lambda < 1$ alter more recent predictions more significantly than previous ones.

TD(0) is another extreme case of TD(λ) algorithm, in which $\lambda = 0$. The method is therefore simplified to:

$$\Delta w = \sum_{t=1}^{m} \Delta w_t, \qquad \Delta w_t = \alpha (P_{t+1} - P_t) \nabla_w P_t. \tag{5.13}$$

In other words, TD(0) algorithm looks only one step ahead and adjusts the estimates based on their immediate difference. This is, again, equivalent to a gradient-based supervised learning procedure, but this time with the value of subsequent prediction P_{t+1} substituted for actual outcome z.

5.3.2 Q-Learning Algorithm

Q-learning [341, 342] is another reinforcement learning algorithm. While the key concept of TD methods was P_t - a value of prediction assigned to each encountered state, Q-learning makes use of $Q(s_t, a)$ function, which assigns value to each state-action pair. This function is expected to approximate the expected reinforcement of taking action a in state s_t. Therefore, $P_t = \max_a Q(s_t, a)$. $Q(s_t, a)$ values approximations can be iteratively improved in a process analogical to the TD-family methods. In its simplest form the learning rule is defined as follows [163]:

$$Q(s_t, a) := Q(s_t, a) + \alpha (r + \gamma \max_{a'} Q(s_{t+1}, a') - Q(s_t, a)), \tag{5.14}$$

where $\alpha \in (0, 1]$ is the learning rate and $\gamma \in (0, 1]$ is the discount factor. The method can easily be extended to incorporate reinforcement values discount factor and update states that occurred more than one step earlier in a way similar to TD(λ) algorithm.

5.3.3   Temporal Difference Learning in Games

The TD(λ) algorithm is often used in game playing to modify weights of the game state evaluation function, which is commonly of the linear form (7.4) presented in chapter 7. This wide applicability stems from the fact that game playing domain is one of the typical environments for TD methods, as it ideally fits the description of an iterative prediction process in which the real final outcome is not known until the whole procedure is finished. Experiments in this field have led to several variations of TD methods, such as TDLeaf(λ), Replace Trace TD or TDMC(λ), and various training schemes (presented in detail in chapter 9).

TD(λ)

Contemporary implementations of the standard TD(λ) algorithm are usually based on a slightly more generalized version of the method. Changes of the weight vector are computed based on the following equation:

$$\Delta w_t = \alpha \cdot \delta_t \cdot e_t \tag{5.15}$$

where $\alpha \in (0, 1)$ is the learning step-size parameter.

The second component $\delta_t = r_{t+1} + \gamma V(s_{t+1}^{(l)}, w) - V(s_t^{(l)}, w)$ represents the *temporal difference* in state values, r_{t+1} is a scalar reward obtained after transition from state s_t to s_{t+1} and $\gamma \in (0, 1]$ is the discount parameter. In many implementations $\gamma = 1$ and $r_t = 0$ for all t, which effectively reduces formula (5.15) to equation (5.12). A small negative value of r_t is sometimes used to promote early wins. $s_t^{(l)}$ is the so-called *principal variation leaf node* obtained after performing a d-ply minimax search starting from state s_t (the state observed by the learning player at time t). In other words $V(s_t^{(l)}, w)$ is the minimax value of state s_t or a d-step look-ahead value of s_t.

The last component of equation (5.15) is the eligibility vector e_t updated in the following recursive equation:

$$e_0 = 0, \qquad e_{t+1} = \nabla_w V_{t+1} + (\gamma\lambda)e_t, \tag{5.16}$$

where $\lambda \in (0, 1)$ is the decay parameter. $\nabla_w V_t$ is the gradient of $V(s_t, w)$ relative to weights w.

The weight updates can be computed either *online*, i.e. after every learning player's move or *off-line*, i.e after the game ends. This latter mode enables to condition weight replacement by the final result of the game (win, loss or draw).

Figure 5.3 presents a pseudo-code for the implementation of TD(λ) learning method in TD-GAC (Temporal Difference Give Away Checkers) system [241]. Initially both eligibility vector and state value estimation are set to 0. After each move a new value estimating the current state is assigned to variable res_2. The difference between estimation of actual state and immediately previous one (*temporal difference*) is assigned to $diff$. Next, the

weights are updated according to (5.15), a new gradient is calculated and the
eligibility vector e is updated according to (5.16). Finally, the actual state
estimation value is moved to res_1 and the algorithm proceeds based on the
next move made in the game.

function TD(λ)
1: $e = [0, 0, \ldots, 0]^T$;
2: $res_1 = 0$;
3: **repeat**
4:      $res_2 = $ VALUE_CURRENT_STATE;
5:      $diff = res_2 - res_1$;
6:      $\Delta w = \alpha \cdot diff \cdot e$;
7:      $w = w + \Delta w$;
8:      $grad = $ GRADIENT_CURRENT_STATE;
9:      $e = grad + \gamma\lambda \cdot e$;
10:     $res_1 = res_2$;
11: **end**;

Fig. 5.3. TD(λ) algorithm

In TD(λ) the weight vector determines the value function which in turn
determines the player's policy $\pi : (S, R^K) \rightarrow A$. Policy $\pi(s, w)$ is a function
that for every state $s \in S$ and some weight vector $w \in R^K$ maps an action
$a \in A$. Thus $\pi(s, w)$ tells the agent which move to perform at state s.

The most natural implementation of policy π it to always perform the move
that follows the best line of play found. The best line of play is determined
by evaluating the leaf nodes of a game tree. Hence, the goal of learning an
optimal policy that maximizes the chance of a win is achieved indirectly by
learning an optimal value function.

Following the greedy policy in each move can lead to insufficient state space
exploration. The alternative, known as an ε-greedy policy, is to introduce
random instabilities in the process of optimal move selection. Based on the
results presented in [213, 242] it may be suggested that in the case a greedy
policy is followed, the state space exploration can be forced by playing in
turn with relatively large number of diverse opponents. This topic is further
discussed in chapter 9.

TDLeaf(λ)

TDLeaf(λ) algorithm [19, 20] is a variant of basic TD(λ) method that enables
it to be used together with the minimax search method. In TDLeaf(λ) gradient at time t (line 8 in figure 5.3) is calculated with respect to the *principal
variation leaf node* $s_t^{(l)}$ instead of state s_t. In other words, gradient in (5.16)
is applied not to the current game state, but to the most promising leaf state
chosen based on the minimax search performed from state s_t.

Replace Trace TD

Another variant of TD-learning method known as *Replace Trace TD* was introduced in [302]. The underlying idea of this method is specific actualization of eligibility vector e.

Note that in classical TD approach the elements of e can theoretically grow or decrease without limits (c.f. equation (5.16)). For example, consider linear evaluation function (7.4) and assume that during a game the agent frequently encounters states, in which a certain feature is present. In such a case, a component of e corresponding to this feature would grow constantly, achieving relatively high value. According to the authors of [302] such "uncontrolled" increase can be disadvantageous, and they proposed the following scheme of updating e (in line 9 of figure 5.3):

$$e_{t+1}^{(i)} = \begin{cases} \frac{\partial V(s_{t+1},w)}{\partial w_i} & \text{if } \frac{\partial V(s_{t+1},w)}{\partial w_i} \neq 0 \\ \lambda e_t^{(i)} & \text{if } \frac{\partial V(s_{t+1},w)}{\partial w_i} = 0 \end{cases} \tag{5.17}$$

where $e_t^{(i)}$ is the ith element of vector e_t. It is clear from equation (5.17) that in Replace Trace TD method the components of e are *replaced* with the actual values of $\frac{\partial V(s,w)}{\partial w_i}$ and not cumulated in time as is the case in the original formulation of TD method.

TDMC(λ)

One of the hurdles in applying TD learning methods in games is connected with the delayed reward which is not known until the end of the game. Therefore, a new algorithm by the name of Temporal Difference with Monte Carlo simulation (TDMC) [240] has been devised. Just as its name suggests, it combines concepts of TD method and Monte Carlo simulations. This goal is achieved by replacing in the learning process the actual output of the games played with winning probabilities obtained via means of simulations. Thanks to this change, partial reward values are available even before the game is finished - what is more, they are independent of previous moves and not susceptible to distortions caused by inaccurate evaluation functions.

TDMC(λ) method requires introduction of several new concepts. First, in order to define the reward value in each position, the return R_t must be defined:

$$R_t = \sum_{i=t}^{m} \gamma^{i-t} r_i. \tag{5.18}$$

In the above equation r_i is a result of running Monte Carlo simulation for position i and γ is a discount rate. The above definition may be simplified to an n-step return $R_t^{(n)}$ by replacing subsequent rewards after $t + n$ with the current evaluation value V_{t+n} (V_t is just a shorthand notation for $V(s_t, w)$):

$$R_t^{(n)} = \sum_{i=t}^{t+n-1} (\gamma^{i-t} r_i) + \gamma^n V_{t+n}. \tag{5.19}$$

Introducing eligibility rate $\lambda \in [0, 1]$ leads to a definition of λ-return R_t^λ:

$$R_t^\lambda = \sum_{n=1}^{m-t-1} (\lambda^{n-1} R_t^{(n)}) + \lambda^{m-t-1} R_t. \qquad (5.20)$$

R_t^λ is defined as target value that should be approximated by evaluation values V_t. Thus, the learning formula is:

$$\Delta w = \alpha \sum_{t=1}^{m-1} (R_t^\lambda - V_t) \nabla_w V_t, \qquad (5.21)$$

where $\alpha \in (0, 1]$ denotes a learning rate.

The TDMC(λ) method was applied to the problem of learning evaluation function for the game of Othello [240]. The learning game records were obtained in a self-play mode, using ε-greedy policy (c.f. equation (9.3) in section 9.1.4). Various values of ε were tested, and in all choices the TDMC(λ) method was superior to the traditional off-line TD(λ) approach. Quite surprisingly, the above advantage includes also the case of $\varepsilon = 1.0$, which essentially means playing a completely random game. According to the authors this phenomenon should be attributed to the way TDMC(λ) assigns the target value in a given position, which is based not on the final outcome of the game, but on the Monte Carlo-based estimated probability of winning calculated in a given nonterminal state.

5.4 Summary

Neural networks, evolutionary methods and reinforcement learning algorithms are the main CI disciplines used in mind board game playing. Contemporary CI-based game playing systems are usually non-homogenous and combine methods and tools from two or sometimes all three of the above paradigms. In majority of such hybrid solutions, the role of neural networks is to implement the evaluation function - in most cases in the form of a linear or nonlinear MLP. The weights are usually trained either through evolutionary process or according to the reinforcement learning method. Evolution is often performed without the use of an external fitness evaluation function, i.e. through direct competition between peers (different candidate solutions). Such a coevolutionary process is widely applicable to a large spectrum of games mainly due to the lack of any requirements concerning the use of additional expert knowledge about the game, other than the game rules and goal states definition and assessment.

Another possibility of neural network weights training is the use of reinforcement learning method. The standard approach in this case is application of one of the several derivatives of the temporal difference method, usually TD(λ), TDLeaf(λ), or Replace Trace TD. Yet another possibility of hybrid approach is the use of neuro-evolutionary system, in which not only the weights but also the neural architecture is evolved, within some predefined range.

In some cases solutions proposed within these three fundamental CI areas are enhanced by applying techniques originating in other soft computing fields, e.g. Bayesian learning [41, 308] or swarm intelligence [115, 244]. Recently, application of Monte Carlo simulations has gained great popularity, especially in Go and in imperfect-information games like poker or bridge.

The choice of a particular method of solving the game related task usually depends on both the task definition and the properties of the candidate solution. In the remaining part of the book, several examples of successful application of neural nets, evolutionary methods and TD-learning systems to mind board games are presented and discussed in the context of the suitability of a given solution method to the task being solved.

Detailed analysis of pros and cons of particular implementations may hopefully show the potential future perspectives of CI-related systems in game domain, but at the same time point out some major impediments in building truly autonomous, self-contained, human-like artificial game players.

6

CI in Games – Selected Approaches

This chapter starts with presentation of two notable examples of CI application to mind game playing: Tesauro's Neurogammon and TD-Gammon systems playing backgammon (chapter 6.1), and Chellapilla and Fogel's Blondie24 approach to checkers (chapter 6.2). Both these systems are milestone accomplishments in CI-based game playing and both had inspired several developments that followed their underlying concepts. Next, in chapters 6.3 and 6.4, respectively a few other examples of TD and neuro-evolutionary learning systems are briefly introduced.

6.1 TD-Gammon and Neurogammon

The first world-class accomplishment in the area of CI application to games was TD-Gammon program [317, 318, 319] and its predecessor - Neurogammon [316]. Both were written by Gerald Tesauro for playing backgammon - an ancient two-player board game. In short, the goal of the game is to move one's checkers from their initial position on one-dimensional track to the final position (players make moves in the opposite directions). The total distance that pieces belonging to one player can move at a given turn depends on the score on two dice rolled by the player at the beginning of each move. Based on the dice' score the player makes a decision regarding which pieces to move forward and how many fields. Due to dice rolling, which introduces randomness into the game, backgammon has a high branching factor (exceeding 400) and when played by masters becomes a highly complex battle, full of tactical and positional threats including sophisticated blocking strategies. Even though the rules of the game are quite simple, expert game strategies are fairly complicated and exploit various game subtleties.

In Neurogammon, the evaluation function was implemented as an MLP network trained with backpropagation, having as an input the location of pieces on the board and a set of game features carefully designed by human experts. Board positions were extracted from the corpus of master-level

J. Mańdziuk: Knowledge-Free and Learning-Based Methods, SCI 276, pp. 71–89.
springerlink.com   © Springer-Verlag Berlin Heidelberg 2010

games and manually labeled with relative values of possible moves. Neurogammon achieved a stable human intermediate level of play which allowed it to convincingly win the 1989 International Computer Olympiad [316].

Quite a different approach was adopted in TD-Gammon - the successor of Neurogammon. TD-Gammon was also utilizing the MLP network, but differed from Neurogammon in three key aspects: (1) temporal difference learning was used instead of backpropagation training; (2) a raw board state without any expert features constituted the input to the network[1]; (3) training was essentially based on self-playing as opposed to training relying on board positions that occurred in expert games.

Initially, when a knowledge-free approach was adopted, TD-Gammon reached an intermediate human level of play, roughly equivalent to Neurogammon's level. In subsequent experiments - still in a self-playing regime, but with the input layer extended by adding expert board features (the ones used by Neurogammon) to the raw board data - its level of play eventually became equivalent to the best world-class human players.

Following Tesauro's work, various attempts to repeat his successful TD approach in other game domains were undertaken, but none of the subsequent trials reached as high level of playing competency in any game as TD-Gammon did in backgammon.

One of possible reasons of TD-Gammon's striking efficiency is stochastic nature of the game which assures a broad search of the entire state space and alleviates the exploration-vs.-exploitation dilemma commonly encountered in TD-learning (see section 9.1.4 for detailed discussion on this subject). Another reason is attributed to using a real-valued, smooth, continuous target evaluation function, for which small changes in game position cause adequately small changes in position assessment. This is generally not the case in most of the popular perfect-information, deterministic games where discrete functions are employed. Yet another reason is ascribed to the impact of TD-learning strategy on the process of neural net's training: simple linear associations were learned first, and only then a representation of nonlinear, context-sensitive concepts and exceptional cases was built [319]. A final remark concerns the observation that in backgammon shallow search is good enough when playing against humans [20], who, due to stochastic nature of the game, use only 1-2-ply search.

Tesauro's experiment proved the capability of MLP networks to learn complex nonlinear associations based on temporal difference method and self-playing. Training method used in TD-Gammon appeared to be clearly superior to the supervised example-based training implemented in Neurogammon.

One of the very intriguing results of the TD-Gammon experiment was the ability of the system to come up with genuinely new ideas concerning positional aspects of the game - this issue is further discussed in chapter 13

---

[1] Some experiments with adding expert features to the input vector were carried out in subsequent studies.

devoted to creativity and automatic feature discovery in game playing systems. Another inspiring observation is the existence of specific patterns in the MLP's weight space which reflect particular backgammon features [317]. Similar phenomena were later discovered by the author of this book in MLP-based realizations of bridge-related experiments [209, 235]. This topic is expanded on in chapter 13.

TD-Gammon was the first successful large-scale attempt of applying the TD-learning in game domain and despite some critical discussions concerning the underlying reasons of this success (being, according to [257], mainly the nature of the game rather than the choice of the learning method) remains a milestone achievement in this area.

6.2   Anaconda vel Blondie24

Another milestone achievement in CI game playing is *Blondie24*, an expert level checkers program. In the late 1990s Kumar Chellapilla and David Fogel carried out a coevolutionary experiment aimed at developing efficient evaluation function for the game of checkers without using *a priori* domain knowledge [3, 62, 63, 64, 65, 66, 109, 111]. An ensemble of feed-forward neural networks, each representing a candidate solution, was evolved in an appropriately designed evolutionary process.

Each network was composed of a 32-element input layer, followed by three hidden layers with $91, 40$, and 10 neurons, respectively and a single-node output layer. In each neuron (except for the input ones) hyperbolic tangent was used as an internal transfer function.

The input data for each network consisted of locations of pieces on a game board represented by a 32-element vector. Each component of the vector could take one of the following values $\{-K, -1, 0, 1, K\}$ for some $K > 1$, where positive elements represented pieces belonging to the player and negative ones - the opponent's pieces. Value of $\pm K$ was assigned to a king and ± 1 was assigned to a regular checker. The value of 0 represented an empty square.

The role of the first hidden layer was spatial preprocessing of the input information divided into 91 local, partly overlapping chunks. In effect the input data was decomposed in the first hidden layer into all possible squares of size 3×3 (there were 36 such squares), 4×4 (25 ones), ..., 8×8 (one square) of the entire board. Each of the above squares was assigned to one first hidden layer neuron (fig. 6.1).

The two subsequent hidden layers operated on features originated in the first hidden layer. A single output neuron represented the evaluation of a position presented in the input. The closer its value was to 1.0 the more favorable position for the player was and the closer the value to -1.0 the more favorable position for the opponent.

Apart from input from the last hidden layer, the output neuron received also the sum of all elements of the input vector representing the board position in question, as additional input (fig. 6.1). Actually, this was the only feature

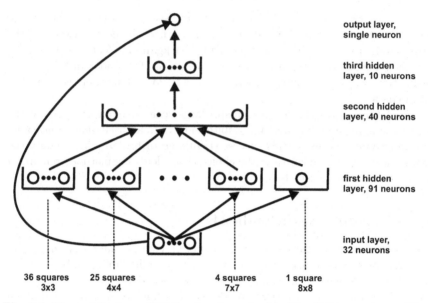

output layer,
single neuron

third hidden
layer, 10 neurons

second hidden
layer, 40 neurons

first hidden
layer, 91 neurons

input layer,
32 neurons

36 squares
3x3

25 squares
4x4

4 squares
7x7

1 square
8x8

Fig. 6.1. The architecture of MLP networks evolved in Blondie24 experiment. The board is represented as a one-dimensional 32-element vector. Each of the first hidden layer neurons is connected only to specific subset of the input elements representing one of the 36 possible (partly overlapping) 3×3 subsquares of the board or one of the 25 4×4 subsquares,..., one of the 4 7×7 subsquares. The last neuron in the first hidden layer is connected to all input elements (covering the entire board). The second and the third hidden layers are fully connected and composed of 40 and 10 elements, respectively. In the single output neuron the estimated evaluation of the checkers position encoded in the input layer is calculated. Additional direct link from the input layer to the output neuron provides the sum of all 32 input elements in order to help the network "sense" the importance of piece count difference

that could be claimed to be linked to expert knowledge. On the other hand the difference between plain sums of pieces' values for both playing sides does not seem to be closely related to the specificity of checkers - it is rather a more general feature, applicable in most of the classical mind board games. Since the value of a king (i.e. $\pm K$) was also evolved, the evolutionary process, by developing a suitable value of K, could potentially override the piece differential feature at least with respect to appropriate weighting between kings and ordinary checkers.

The evolutionary population consisted of 15 individuals $P_i, i = 1, \ldots, 15$, each of them defined by a variable value of a king K_i and a vector of weights and biases related to the above described (fixed) neural network architecture. For each individual P_i, the value of K_i was initially set to 2.0 and weights and biases were randomly sampled from uniform distribution over $[-0.2, 0.2]$. For each P_i, mutation operator was controlled by a vector of self-adaptive parameters σ_i corresponding one-to-one to weights and biases of the network and restricting the mutation search space. All elements of $\sigma_i, i = 1, \ldots, 15$ were initially set to 0.05.

In each generation, each parent network P_i generated an offspring network by mutating weights, biases and possibly king's value. Specifically an offspring P_i' was created from P_i in the following way:

$$\sigma_i'[j] = \sigma_i[j] \cdot exp^{\tau \cdot N_j(0,1)}, \quad j = 1, \ldots, N_{wb} \tag{6.1}$$

$$w_i'[j] = w_i[j] + \sigma_i'[j] \cdot N_j(0,1), \quad j = 1, \ldots, N_{wb} \tag{6.2}$$

$$K_i' = K_i + d, \quad i = 1, \ldots, 15 \tag{6.3}$$

where N_{wb} is the number of weights and biases in the network (equal to 1741), $\tau = \frac{1}{\sqrt{2\sqrt{N_{wb}}}} = 0.1095$ and $N_j(0,1)$ is standard Gaussian distribution with zero mean and variation equal to one, sampled independently for each $j = 1, \ldots, N_{wb}$ and each $i = 1, \ldots, 15$. King's update parameter d was chosen uniformly at random from the set $\{-0.1, 0, +0.1\}$. The value of K_i' was further constrained to the interval $[1.0, 3.0]$ by the following rule:

$$\text{if } K_i' > 3.0 \text{ then } K_i' := 3.0 \text{ else if } K_i' < 1.0 \text{ then } K_i' := 1.0 \tag{6.4}$$

After the above described mutation each of the 30 networks (parents and offspring) played against five randomly selected opponents (with replacement) from that population. The network being evaluated played red all five games. Note that under the above rules the numbers of games played by the networks were in general not pairwise equal (though each of them participated in 10 games, on average).

In order to promote the "not to lose" attitude of the evolved players the payoff of the game was not a zero-sum, but was biased by assigning $1, 0, -2$ points for win, draw and loss, respectively. The above protocol was applied to both the evaluated player and its opponent. A draw was declared after 100 moves were played by each side. The games were played using alpha-beta with search depth $d = 4$ ply. This depth was extended by two whenever a forced move (a unique possible move in a given position) was encountered. The search was also extended by two in case the finally reached position was not stable (involved a jump move). After all 150 games were played the top-15 networks constituted the population for the next generation. After 230 generations the evolution was stopped and the top network was tested against human competitors on the Internet site, where it played 100 games, generally with depth $d = 6$ and occasionally with $d = 8$ [62]. 49 out of 100 games were played with red pieces.

The network achieved the rating of an A-class player (immediately below the expert level) according to some approximate measurement procedure [62]. After continuation of the coevolutionary process described above, the best network developed in the 840th generation achieved the rating of an expert player (just below the master level) of about 2045 ELO points, playing with search depth of $d = 8$ on the same Internet gaming site [65, 66, 109].

This network was also tested against three characters (Beatrice, Natasha, and Leopold) from the *Hoyle's Classic Games* - commercially available software - winning a six game match with the score 6 : 0 [64].

The final assessment was made by confronting Blondie24 with Chinook - The Man-Machine World Checkers Champion at that time and currently unbeatable program described in chapter 4.2. In the head-to-head competition Chinook was playing on a reduced level equivalent to high expert with roughly $2150 - 2175$ ELO ranking points. In a 10-game match with swapping sides after each game Chellapilla and Fogel's program accomplished two wins with four draws and four losses [111]. Such a result is approximately rated at $2030 - 2055$ ELO points which correlates with previous estimations.

The authors used two names for the evolved player: Anaconda and Blondie24. The former one was related to the system's style of playing which mainly involved restricting the opponent's mobility, while the latter - to attract other players' attention on the Internet gaming zone by pretending to be an attractive 24-year old female player.

It is worth noting that preliminary attempts to coevolve a neural checkers player *without* the spatially-aware first hidden layer (i.e. using only the two fully connected hidden layers of sizes 40 and 10, respectively) presented in [63] were less successful than the preliminary results attained *with* the use of this additional layer [62]. Both approaches are straightforwardly comparable, since except for the existence or not of the spatial hidden layer, they differed only by the definition of the king's mutation rule (6.3) which in [63] was of the following form:

$$K'_i = K_i \cdot exp^{\frac{1}{\sqrt{2}} \cdot N_i(0,1)}, \quad i = 1, \ldots, 15 \tag{6.5}$$

where $N_i(0,1)$ was a standard Gaussian distribution sampled independently for each $i = 1, \ldots, 15$. The resulting value was restricted to interval $[1.0, 3.0]$ according to (6.4).

Comparison of these results suggests that the notion of local spatial arrangement of checkers is very helpful in developing the internal network's representation of the board position, hence allowing its more accurate assessment.

A salient aspect of Chellapilla and Fogel's development is the use of almost completely knowledge-free approach. Only basic technical knowledge, which is given to any novice player was provided to the network. This includes location of pieces on the board, the rules and the goal of the game, and the importance of the difference between checkers possessed by the player and by his opponent. The only expert knowledge can, to some extent, be attributed to the design of neural network architecture, which stresses the importance of local patterns in checkers by dividing the board into overlapping squares of various sizes. On the other hand, the weights reflecting the relative importance of this local information were developed by the network without human supervision. Moreover, the network did not receive the feedback about its performance in individual games (which games were won or lost or drawn), only the total score was provided.

Challapilla and Fogel's experiment was reproduced with some minor differences by Hughes, who reported the results during CIG'05 tutorial [157]. Similar neural architecture was used to represent the evaluation function, which was coevolved for same 840 generations with a population composed of 15 individuals. Hughes pointed out three differences compared to Blondie24' evolution: play at red/white side was random and not biased to red; there was small amount of randomness added to the evaluation network's output; and no search extension was applied when final move was a capture. Each of the above listed differences might potentially have an important impact on the final result, but the detailed analysis of this issue was not published (to the author's knowledge).

The best evolved player called *Brunette24* was pitted against Blondie24 in a two-game match. Brunette24 lost narrowly when playing white and drew (having a king advantage) when colors were reversed. Brunette24 was also successfully playing on the MSN gaming zone, with a 6-ply search depth, but the number of games was too small to allow making reliable conclusions [157].

One of the main conclusions drawn by Hughes was the importance of the *bootstrapping* effect of piece difference value. Setting the respective gene to zero resulted in poor gradient and caused the problems with starting the evolution. Recall that the piece-difference weight was linked directly to the output node (c.f. figure 6.1).

The idea of autonomous development of the evaluation function in a coevolutionary process was very influential and several approaches to other games followed Blondie24 project. Two examples are presented below in chapter 6.4.

6.3 TD-Learning Systems

Tesauro's work on backgammon brought about wide research opportunities in learning computer playing policy in games. Following his successful work, the TD-learning was applied by other authors to checkers [288], chess [19], Go [275, 295], Othello [199, 338], and other games [165, 174]. Three examples of those attempts, respectively in chess, checkers, and give-away-checkers are briefly presented in the following sections.

6.3.1 KnightCap

KnightCap, a chess program written by Baxter, Tridgell and Weaver [18, 19], is one of well-known examples of TD-type learning in games. The authors applied the TDLeaf(λ) method[2] and on the contrary to several previous works [24, 277, 317, 323] found self-playing to be a poor way of learning

---

[2] Although the idea of TDLeaf(λ) was first presented in [24], the algorithm's name was coined in Baxter et al.'s papers.

and preferred using external trainers instead. Hence, the TDLeaf(λ) learning was carried out by playing on the Internet chess site. The program started from the blitz rating of 1650 ELO points and after only three days of playing (308 games) reached the blitz rating of 2150 points, which is roughly equivalent to master candidate player. Afterwards the rating curve entered a plateau.

The success of KnightCap laid, according to the authors, in appropriate choice of TD-learning parameters, and first of all in "intelligent material parameters initialization". The initial estimations of particular pieces' values reflected the common chess knowledge: value of 1 for pawns, 4 for knights and bishops, 6 for rooks and 12 for a queen[3]. Such "reasonable" initialization already located the KnightCap's evaluation function in feasible area of the parameter space allowing its fast improvement.

Another contribution to rapid rating increase was attributed to the fact that the weights of all other (i.e. nonmaterial) parameters were initially set to zero, and therefore even small changes in their values caused a relatively significant increase in the quality of play.

The weakest feature of KnightCap, which hindered its further improvement, was poor play in the opening phase[4]. After implementation of the openings learning algorithm and further development of the evaluation function, the system increased its performance to the level of 2400 – 2500 ELO with a peak value at 2575 points [20]. This version of KnightCap could easily beat *Crafty* [158] - the strongest publicly available freeware chess program and a direct descendant of a former Computer Chess Champion - Cray Blitz [159], if both programs searched to the same depth.

KnightCap's evaluation function was of linear form, composed of 1468 features with coefficients automatically tuned by the system. Four collections of these coefficients were used, respectively for the opening, mid-game, endgame, and mating phases.

KnightCap was applying the MTD(f) search algorithm supported by several heuristic improvements: null moves, selective search extensions, transposition tables, and other. Special care was devoted to efficient move pre-ordering in the searched tree, where a combination of killer, history, refutation table, and transposition table heuristics was applied.

Three main lessons from KnightCap's experiment are of general importance. First, it was shown that appropriate choice of initial weights in the evaluation function is crucial to the speed and quality of training. Second, it is important that the training opponents are comparable in playing strength to the learning program. This observation is in line with common human intuition that too strong or too weak opponents are not as valuable as the

---

[3] Another popular choice is to assign the values of 1, 3, 5, 9, respectively to pawns, knights and bishops, rooks, and a queen.

[4] Due to the way TD-learning is performed the relatively weaker play in the openings is common to practically all TD implementations regardless of the game chosen.

ones playing on approximately the same level as the learner. Since people are generally willing to challenge the opponents of comparable playing strength as themselves, playing on the Internet gaming zone allowed the program to "grow up" and play with opponents of gradually increasing strength. Third, the authors underlined the importance of playing against external opponents versus a self-play learning mode. The pros and cons of both training schemes are analyzed in chapter 9.1. The diversity of opponents is particularly important in the context of alleviating the exploration-vs.-exploitation tradeoff.

6.3.2 TDL-Chinook

Jonathan Schaeffer, the main contributor to Chinook program described in chapter 4.2, together with Markian Hlynka and Vili Jussila applied TD-learning to checkers [288] in order to verify its efficacy in another (after backgammon and chess) demanding game. The authors used the TDLeaf(λ) learning scheme. Their direct goal was comparison between Chinook's evaluation function and the TD-based learned one. In order to allow such a comparison, TD-learning player was initially equipped with Chinook's evaluation function but with a different set of weights assigned to its components. Four sets of 23 coefficients, each related to a particular game phase, were used. Since the values of an ordinary checker and a king were not modifiable, altogether there were 84 tunable coefficients to be set.

Two main approaches were considered: in the first one, Chinook served as the training opponent for the TD player whereas the second approach relied on self-play. In the first case training was performed according to predefined scenario, which involved the use of 144 standard checkers openings, each 3-ply long. Afterwards the game was continued by the programs to the final completion. In both cases separate experiments were run for search depths of 5, 9, and 13 ply, in each case generating two weight sets, for red and white, respectively.

Surprisingly enough, it turned out that in the case of training with teacher by applying the TDLeaf(λ) learning scheme, the program was capable of reaching the level of play comparable to the teacher's, even though Chinook evaluation function weights had been carefully tuned for more than five years. Even more surprisingly, the other approach (self-playing) also led to a Chinook caliber program. The success of a self-play learning implies that external teacher is not indispensable for achieving human championship level of play in as complex game as checkers! This statement is evidently at variance with the Baxter et al.'s conclusions. The most probable reason for this discrepancy was much smaller number of coefficients to be tuned in case of TDL Chinook compared to KnightCap, which made the former task easier.

Detailed analysis of the efficacy of training performed with various search depths confirmed Schaeffer et al.'s conclusion that "There is no free lunch; you can't use shallow search results to approximate deep results" and therefore "the weights [of an evaluation function] must be trained using depths of search expected to be seen in practice" [288].

Another general observation concerns the differences between the weights developed by TDL Chinook and the hand-tuned Chinook weights. Closer examination revealed several interesting insights into how some "human-type" features, which very rarely occur in machine-versus-machine play, are compensated by other components in the evaluation function. Nevertheless it seems reasonable to conclude that complementary training with human master players would be desirable in order to precisely tune the weights assigned to these infrequent features.

6.3.3   TD-GAC

TD-GAC experiment [213, 241, 242] aimed at applying TD-learning methods to give-away checkers (GAC). The game of GAC is played according to the same rules as checkers, differing only in the ultimate goal. In order to win a game of GAC player has to lose all his checkers or be unable to perform a valid move [4]. Despite the simplicity of the rules, GAC is not a trivial game and application of simple greedy strategy of losing pieces as quickly as possible leads to unsatisfactory performance. One of the reasons for unsuitability of the greedy policy is the observation that a single piece is barely mobile. Fig. 6.2 presents two situations in which white loses despite having only one piece left.

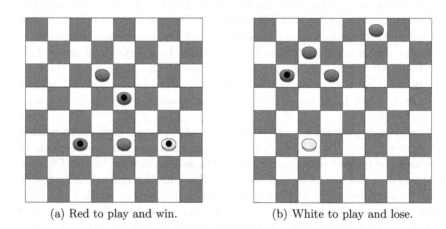

(a) Red to play and win.   (b) White to play and lose.

Fig. 6.2. Examples of inefficiency of greedy heuristics. White move bottom-up. Checkers with small black circles denote kings

The main idea of TD-GAC experiment was to check if a computer player could be trained to reasonable level of play using simple, linear value function and the TD algorithm. Both basic TD approaches: TD(λ) and TDLeaf(λ) were tested and compared. Additionally TD methods were confronted with EVO algorithm, a simple coevolutionary method described in chapter 9.3.

Following [277] linear evaluation function (7.4) with $n = 22$ feature detectors was used. Each feature represented a specific numerical property of the game state, for example the number of pieces on the board, the number of pieces on the diagonal, the number of some predefined formations, etc. Every feature was calculated as a *difference* between the respective figures for the player and the opponent. For example, a piece differential term was used instead of two separate features describing the number of checkers on each side.

The evaluation function was used to assess game positions in the leaf nodes in a 4-ply minimax game tree search. After that, a move on the principal variation path was executed.

Training was divided into three stages, each composed of 10,000 games. A computer player had its weights changed after each game. At each stage 10 players were trained, each of them with a set of 25 opponents/trainers. Trainers played games in a fixed order according to a randomly selected permutation. In the first stage all weights of players 1 and 6 were initialized with zeros. The remaining players had their weights initialized with random numbers from the interval $(-10.0, +10.0)$.

The first stage was the only one, in which pairwise different trainers for each learning player were used. Each of 250 trainers had its weight vector initialized randomly within interval $(-10.0, +10.0)$. Weight vectors did not change during learning. Players from 1 to 5 always played red pieces (they made the first move). Players from 6 to 10 played white pieces. In each game the winning side received 1 point, the losing one 0 points, and both players scored 0.5 point in case of a draw.

The second stage was organized in a very similar way except for two main differences: the trainers were shared among the learning agents and most of them were no longer random players. Twenty weight vectors used by the trainers were selected among the best learning players obtained at different time moments in the first training phase. The remaining 5 vectors were initialized randomly.

In the third stage a similar trainer enhancement process was performed, however, this time all 25 trainers used weights of the most successful players developed at different time moments during the second training stage.

In order to verify whether the performance in the training phase really reflects policy improvement, test phases were introduced during each training stage after every 250 games. In the test phase current weight vectors of the 10 learning players were pitted against unknown testing opponents not encountered earlier in the learning process. Two groups, each composed of 100 benchmark players, were used. One group consisted of random players, the other one - of strong opponents trained previously in another experiment. Learning mechanism was disabled during test games.

Generally speaking, TD(λ) and TDLeaf(λ) learning methods were both effective in achieving an above-average level of play. The ratio of games won against both random test opponents and strong, skillful ones was gradually increasing. After 20,000 training games the score of the best learning agent exceeded 75% against random players (the details are presented in chapter 9.3).

It is worth noting that the design of the training process as well as the predefined learning mechanisms were very simple: only 22 plain game features were used in the linear evaluation function; game tree search depth was equal to 4; the same set of features was used for the entire game; no opening book or endgame database was involved in the process. Despite this simplicity, stable progress of the TD training was observed, which clearly outperformed the simple coevolutionary algorithm presented in [174].

A gradual increase in strength of the trainers used during subsequent stages of the experiment proved to be crucial for the quality of the finally generated playing agents.

Besides the above general conclusion several specific comments concerning the impact of using random versus skilled opponents on the quality of the training process were drawn in [213, 241, 242]. These observations are discussed in detail in chapter 9 dedicated to effective TD training schemes.

6.4   Neuro-evolutionary Approaches

The idea of developing an evaluation function represented by a spatially-aware MLP network in a coevolutionary process proposed by Chellapilla and Fogel, described in chapter 6.2, was subsequently applied in other domains. Two examples of such applications, referring respectively to Othello and chess are presented below.

The third example of neuro-evolutionary approach discussed in this chapter makes use of a completely different protocol, allowing for flexible definition of neural architecture by applying a specially designed chromosome coding. The fitness of individuals is, in this case, calculated based on results of test games against external opponents.

6.4.1   Blondie24-Like Approach to Othello

Chong et al. [70] closely followed the Blondie24 direction with respect to Othello. The coevolutionary process was designed in exactly the same way as in [62, 66]. The evolved neural networks were MLPs with three hidden layers of sizes $91, 40, 10$, respectively. Connections between the input and the first hidden layer were spatially arranged to represent local information restricted to particular sections of the board (all squares 3×3, 4×4, ..., 8×8).

The minor differences compared to [62, 66] were related to board representation - the input vector was composed of 64 elements, since in Othello all squares of the board are used in play. The search depth of the minimax algorithm when comparing the networks in the evolutionary process was restricted to 2. Population size was smaller than in Chellapilla and Fogel's experiment: the number of parent networks P_i in each generation was equal to 10. Consequently, in each iteration 10 offspring (one from each parent) were generated by mutation of weights and biases vectors $\sigma_i, i = 1, \ldots, 10$. The procedure followed equations (6.1) - (6.3) with $i = 10$, $N_{wb} = 5\,900$ and $\tau = 0.08068$. Since there are no kings in Othello, the equation (6.4) was not used.

In the tournament phase each of the 20 networks (10 parents and 10 offspring) played 5 games (as black) against opponents drawn uniformly from the current population. This nondeterministic comparison procedure was very important for maintaining diversity in the population - occasionally, also weaker individuals had a chance to be promoted due to "lucky" opponent selection. In order to encourage an offensive style of play, in the tournament phase each network received $5, 1, 0$ points for win, draw and loss, respectively (Chellapilla and Fogel used a scale of $1, 0$ and -2).

The evolutionary process lasted for 1 000 generations. In the meantime, populations were tested against two computer players (external opponents) in order to monitor the progress of evolution. The results of these intermediate tests were not provided to the evolved networks nor were they used to improve the evolutionary process. One of the computer players used material-based evaluation function, the other one applied a positional-based evaluation adopted from Rosenbloom's Iago - historically the first master-level Othello program [270]. Playing strength of the tested networks was limited by shallow search depth of 2 ply only. Additionally in all test games the networks played black (made the first move) which, according to the authors of [70], was a significant handicap since computer programs are better in playing white.

Testing the networks during the evolution allowed direct observation of their progress both on the entire population level and individually. Generally speaking, the networks were able to gradually improve their performance against both piece-differential-based and positional-based players. During 1 000 generations they managed to compensate the handicap of playing black as well as outperform the computer opponent that was playing at depth level of 6. Quite interestingly up to about 800 generations the networks were capable of effective play against either one or the other of the computer players but not both of them. Only in the last 200 generations were they able to develop skills allowing them to confront both types of opponents (material and positional). Analysis of the games played by the networks revealed that in the evolutionary process they were capable of discovering the importance of mobility as well as several positional features.

The main conclusions from [70] are in line with remarks presented in Chellapilla and Fogel's papers. High performance of the evolved networks is attributed mainly to the flexibility of the design of the evolutionary process (potential mutation of all weights and biases), spatial representation of the game board in the first hidden layer (local focus on the subsections of a board position) and diversity of the population maintained by using a nondeterministic fitness procedure that allows occasional survival of objectively weaker networks.

6.4.2   Blondie25

Following successful application of coevolutionary approach to checkers David
Fogel and his collaborators verified the efficacy of adopting a similar protocol
to chess - a more demanding and more challenging domain [112]. In this case,
however, the implementation of the method was quite deeply modified, in
particular by incorporating some domain knowledge.

The evaluation function was a linear combination of the following three
factors: (1) sums of the material values of the pieces belonging to each player,
(2) pre-computed tables defining the strength/weakness of having a particular
piece on a particular square - the so-called *positional value tables* (PVTs) -
such tables were defined separately for each type of chess pieces, and (3)
outputs of three neural networks each of which focused on a particular area
of the chessboard (the two first rows, the two back rows, and the 16 squares
located in the center, respectively).

The chessboard position was represented by a one-dimensional vector of
64 elements taking values from the set $\{-K, -Q, -R, -B, -N, -P, 0,
+P, +N, +B, +R, +Q, +K\}$. The element equal to zero denoted an empty
square, the other values were assigned respectively to king, queen, rook,
bishop, knight, and pawn, where positive values represented player's pieces
and negative ones - pieces belonging to the opponent. Initial values for particular pieces were chosen arbitrarily based on the statistical estimation of their
strength (or "common wisdom" - often used by novice players when assessing
the potential material exchange) in the following way: $P = 1$, $N = B = 3$,
$R = 5$, $Q = 9$ and $K = 10\,000$. The value assigned to king reflected the property that a king was "priceless" and couldn't be captured. All these values,
but the one assigned to king, were mutable.

The initial values of PVTs ranged from -50 to $+80$ depending on the piece
and its position on the board (e.g. PVT values for a king were between -50
and $+40$ whereas for pawns between -3 and $+5$)[5]. These values were selected
based on open-source chess programs - hence reflected the expert knowledge
about the game.

The three neural networks were implemented as fully connected MLPs
with 16 inputs, 10 hidden nodes and one output. All weights and biases
were initialized randomly according to uniform distribution from interval
$[-0.025, 0.025]$. Hidden neurons implemented a sigmoid transfer function:
$f(x) = 1/(1 + exp(-x))$, where x represented a weighted sum of all input
signals. In the output nodes also a sigmoid transfer function was used, scaled
to the range $[-50, 50]$ in order for the output values to be comparable to the
values of PVTs.

In the evolutionary process each of 10 parent individuals created one offspring by mutating all variable parameters (material values for the pieces,

---

[5] For each king three PVTs were used depending on whether or not the king
had already castled, and if so whether it had castled on the kingside or on the
queenside.

PVTs' elements, and weights and biases of the three neural networks) - in a similar way as in the case of checkers and Othello using self-adaptive step size. Population of parents and offspring competed for survival by playing 10 games each with randomly selected opponents from the population (excluding themselves). Five games were played with white and five with black color. The players received 1 point for a win, -1 for a loss and 0 for a draw. Search depth d was set to 4 and extended in case of forced moves and whenever the final position was not quiescent. After the tournament, the 10 best players were promoted to the next generation. The experiment was conducted in 10 independent trials, each time for 50 generations. The best overall player was named *Blondie25* and selected for subsequent tests against other computer programs.

Blondie25 played a 120-game match against *Chessmaster 8000* which was playing at the depth level of 6 ply, half of the games with white and half with black pieces. The level of Blondie25's play was estimated at 2437 ELO points. The baseline, non-evolved player using the piece material estimation and PVTs with their initial values (with no use of the three neural networks) was estimated at the level of 2066 points. Hence, the increase of playing strength stemming from applying the evolutionary process was approximately equal to 371 rating points.

Furthermore, Blondie25 was confronted with *Pocket Fritz 2.0*, a popular commercially available chess program, playing 12 games (6 as white and 6 as black) under tournament time restrictions. The evolved system won 9 games, lost 2, and drew 1, achieving the so-called "performance rating" at the level of 2550 points [112]. After continuation of the evolutionary process, the best player obtained in 7462nd generation (when the process was terminated due to a power failure) was again pitted against Pocket Fritz 2.0 achieving performance rating of approximately 2650 points in a 16-game series which resulted in 13 wins, 0 losses, and 3 draws in favor of evolutionary program [113].

In all the above-mentioned contests the evolved player was not implementing any time management procedures that would allow optimization of time usage. Such heuristic procedures were developed later and discussed in [114] along with description of the final assessment of Blondie25 in a match with *Fritz 8.0* (the 5th strongest chess playing program in the world at that time) and with a nationally ranked human chess master (one of the coauthors of [112, 113, 114]). Taking advantage of time management procedures allowed achieving the approximate performance ranking of 2635 points in a 24-game match against Fritz 8.0 (half of the games Blondie25 played with white pieces) with 3 wins, 10 losses, and 11 draws.

Another verification of Fogel et al. program's competency was a 4-game match with a nationally ranked human chess master. Blondie25 won 3 games and lost the remaining one roughly achieving a ranking of 2501 points. This result, attained against a competent human player is particularly interesting since the way of playing chess by humans differs from that of computers. Humans do significantly fewer calculations and play in a more intuitive way than machines, which, on the contrary, are capable of performing extremely fast, deep calculations of possible lines of play, but at the same time lack human-specific cognitive skills.

Certainly, as concluded by the authors, achieving the performance ranking of about 2650 ELO is the highest level ever attained by a program that learned to play chess with the use of internal mechanisms without direct feedback.

On the other hand it should be noted that the design of Blondie25 only in part followed the principles underlying the Blondie24 experiment with knowledge-free approach to playing checkers. In particular Blondie25 used opening and endgame databases designed by chess experts. Also the importance of the center region and the two border rows on each side of the chessboard implicitly incorporated expert domain knowledge into the learning process. Similarly, the PVTs initialized according to human knowledge as well as initial piece estimations were strongly domain knowledge dependent. The estimated performance of a baseline, non-evolved player (which did not take into account the outputs of three neural networks focusing on particular regions of the board) was roughly equivalent to a strong candidate master player. This confirms that starting point of the coevolutionary process was relatively high in the ranking scale.

A fascinating question is whether it is possible to develop a high quality chess player (comparable to a grandmaster level) in a totally "blind," knowledge-free coevolutionary (or other autonomous) development process. This question is yet to be answered.

6.4.3   Moriarty and Miikkulainen's Othello Player

Quite a different approach to neuro-evolutionary learning in games was proposed in [231] for Othello. The networks used in the evolutionary process were not equipped with any "hand-coded" human knowledge about the game, nor any search mechanism was implemented.

Unlike in the Blondie-type approach, where the neural architecture was fixed, the method proposed in [231] relied on non-homogenous population. The topology of resulting neural networks varied between individuals. Sizes and roles of the input and output layers were the only common features. The input layer was composed of 128 units - two units per square. Each unit pair coded a player's piece, an opponent's piece, or an empty square, respectively by the following combinations: $(1,0), (0,1), (0,0)$. A combination of both units being on, i.e. $(1,1)$ was prohibited. Each network contained 64-unit output layer indicating the preferences of moving on particular squares, one output unit per square. The remaining topology and weights of each network's architecture were coded as chromosomes in the form of a string composed of $U2$-type 8-bit integers (ranging from -128 to 127) and were free to evolve.

The variable part of the network architecture was coded according to *marker-based scheme* originally proposed in [119], inspired by biological structure of DNA strands. In short, in the strings coding the network, the hidden nodes' descriptions (definitions of weights) are separated by start-end markers. In particular in [231] the start marker is defined as the value S such that $S \equiv 1 \, (mod \, 25)$ whereas the end marker E fulfils the condition $E \equiv 2 \, (mod \, 25)$. Any integer value between such two neighboring markers is recognized as part of the network's coding.

Schematically, each hidden node with associated weights is encoded in the following way:

$$< S >< node >< in\_val >< key_1 >< label_1 >< w_1 > \ldots$$
$$< key_n >< label_n >< w_n >< E > \qquad (6.6)$$

where $< node >$ and $< in\_val >$ denote the node's label and initial value, respectively, and each of the triples $< key_i >< label_i >< w_i >$ for $i = 1, \ldots, n$ defines particular connection associated with node $node$. Value $key_i > 0$ specifies a connection either from the input node (if $label_i > 0$) or to the output node (if $label_i < 0$). If key_i is negative the label specifies a connection from another hidden unit. For connections from the input layer or the ones to the output layer the respective input or output node is defined as the value of $abs(label)(mod \, N)$ where $abs(x)$ denotes the absolute value of x and N is the number of nodes in the layer. $N = 128(64)$ for the input (output) layer, respectively. The connection within the hidden layers is defined from the hidden unit whose label is closest to the connection's label value. If there is more than one possibility, the first one of them, according to the order in the chromosome's representation, is chosen. The chromosome is "wrapped around", i.e. continues from the end of the chromosome to the beginning until the E marker is encountered (see figure 6.3 as an example).

The network operates on hidden units by activating them one at a time according to the order specified by their location in the chromosome. Hidden units are allowed to retain their values after activation, which can potentially be used as network's short-term memory [231].

According to the authors, the main advantage of using the marker-based coding lies in its natural ability to evolve not only the network's weights (which is the most popular approach to neuro-evolution), but simultaneously also the network's architecture and topology of connections. Another advantage is partial independence of the allele's interpretation of its location in a chromosome, since the order of alleles is not strictly corresponding to the topology of network's weights. Hence, the coding potentially allows the evolutionary process to group relevant nodes (connections) into useful local blocks to be propagated to the next generation.

In the evolutionary process a population of 50 chromosomes, each composed of 5 000 integers, was used. In each generation the top-12 of them were selected for a two-point crossover operation. A mating network for each of them was selected at random from the chromosomes with greater or equal fitness. Two offspring were generated from each of the chosen chromosomes. The resulting 24 networks replaced the worst ones in the population. Mutation was implemented by adding a random value to an allele. The resulting

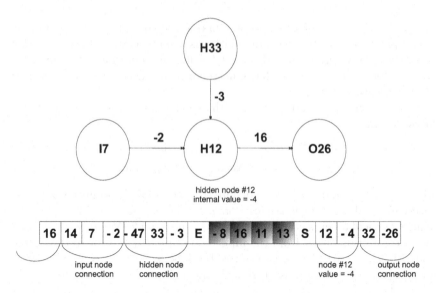

Fig. 6.3. Example of *marker-based coding scheme*. The chromosome encodes a hidden node number 12 (the first value after the S marker). Its internal value is equal to -4 (the next allele). Subsequent alleles describe its connections. For example the connection defined as $< 14 >, < 7 >, < -2 >$ is from the input node (since both 14 and 7 are greater than 0) with connection weight equal to -2. The coding starts with the S marker and ends with the E marker, "wrapping around" the chromosome. Alleles between the E and S are not used. In the case of more complex neural structures the chromosome would consist of many marker pairs $< S > \ldots < E >$ each devoted to one particular hidden neuron and its connections

value of the allele was "wrapped around" if it exceeded the valid interval of $[-128, 127]$. Mutation probability was set to 0.4%.

The fitness function was based on the number of games won out of 10 played by the network against random player in the initial phase of the evolution, and against an alpha-beta search-based opponent ($\alpha\beta$-player) searching with depth $d = 3$, in the subsequent phase. The $\alpha\beta$-player employed positional strategy similar to that of Iago [270]. A complex mobility strategy being part of Iago's evaluation function was not implemented in order to give the networks a chance to exploit potential mobility weaknesses.

In the first phase, when playing against random move generator the networks immediately started to develop positional strategies which were mature enough to allow them to surpass the opponent by winning 97% of games after only 100 generations. Closer examination of individual games revealed that in most of the games both players monotonically increased their piece counts in the initial phase of the game accomplishing similar scores. However, due to stronger positional play of the network, at some point in the game, it visibly started to increase its piece count advantage. At stated by the authors: "The network's strategically anchored positions allow efficient capture of the random mover's pieces without the need to sacrifice many of the network's own pieces."

After initial 100 generations, the random opponent was replaced by a more demanding $\alpha\beta$-player. Despite initial period of poor behavior that occurred right after this switch, the networks' skills were gradually increasing and after 2 000 generations the best evolved network was capable of winning about 70% of the games against the $\alpha\beta$-trainer. More importantly, a closer look at the games played by the networks uncovered a change in their playing style which evolved from purely positional to more advanced mobility-based strategy. The networks "were not as interested in taking edge pieces as before. Instead, their attention was focused on moves that captured fewer discs and controlled the center of the board position. However, the value attached to corner squares had not changed, and if presented a legal corner move the networks would immediately capture it." The above excerpt from [231] confirms that the ultimate strategy picked up by the evolving networks was a mixture of mobility and positional factors with the former being the prevailing ones.

Analysis of games proved that in the evolutionary process the networks were able to autonomously discover the importance of mobility, even though its relevance was neither explicitly pointed out nor underlined in any way during the learning process. It is worth noting that mobility strategies in Othello are counterintuitive and as such generally difficult to understand and develop by novice human players. Human intuition suggests a greedy strategy relying on capturing as many pieces and conquering as large territory as possible.

Moriarty and Miikkulainen's approach to Othello was one of the very first successful examples of search-free learning without the use of domain knowledge. Its success, according to the authors, should be attributed to a large extent to marker-based encoding scheme.

In the course of training the networks showed a great potential to adapt to the new playing style when the training opponent was changed from random to $\alpha\beta$-player. Having a more demanding opponent the networks were able to find the way to confront this challenge by discovering the importance of mobility, the use of which was not necessary when playing against the random player. The mobility feature was realized by keeping the number of network's own disks relatively low and instead paying greater attention to controlling the center of the board and retaining more move options.

Part III

An Overview of Challenges and Open Problems

Evaluation Function Learning

The common aspect of CI game-playing systems is their ability to gradually improve their playing strength, which is mainly realized through learning from experience based on the games played.

In majority of the cases learning is implemented in one of three following ways: as a neural network training, in the form of an evolutionary process, or as a reinforcement learning system, as discussed in chapter 5. Regardless of the learning method used, and the choice of target game, the ultimate goal is development of a high-quality, close-to-optimal evaluation function. The quality of position evaluation together with the efficacy of the search method in most cases determines the overall strength of the playing system. In the world-class playing programs both aspects are painstakingly addressed and both highly contribute to the final program's rating. In the CI motivated approaches the focus is much more on the evaluation function aspect rather than the search, since the latter - in its technical implementation - is located outside the scope of CI.

Learning an evaluation function for a given game can be regarded in two general contexts. First, there are methods which rely on predefined sets of game features and focus on appropriate - linear or nonlinear - combination of them. Methods of this type are of interest in this chapter. The other, more sophisticated possibility is to implement some *creativity* or *knowledge discovery* mechanisms, that may lead to autonomous finding of - previously unknown - relevant game features or elements of game representation. These methods, discussed in chapters 13 and 14, are particularly challenging for CI.

In the remainder of this chapter it is assumed that the set of game (input) features is predefined (usually by human experts or based on analysis of a large number of game positions performed prior to the learning process) and remains unchanged during the entire learning. Given this assumption, there are three major factors, which can be identified as being crucial to the success of the evaluation function learning process. These are: the choice of the features, the form of the evaluation function, and the choice of the learning method and its parameters.

J. Mańdziuk: Knowledge-Free and Learning-Based Methods, SCI 276, pp. 93–97.
springerlink.com © Springer-Verlag Berlin Heidelberg 2010

The Set of Features

The basic factor contributing to efficient learning process is the choice of features used to describe game states. On the one hand a selected feature set should be as compact as possible (since the more features the more weights to be tuned), but at the same time should allow representation of all relevant game aspects and various position peculiarities. Thus, designing the optimal (in size and representational capabilities) set of game features is a highly nontrivial task with huge impact on final quality of the evaluation function developed. Certainly, when designed by human experts, the set of features is usually game-dependent and reflects the human knowledge about the game. In practice it is common that a few sets of features, each dedicated to particular game phase, are designed for the playing agent.

Quite a different protocol is adopted in purely knowledge-free approaches, in which typically only a raw description of the game state is provided to the system. In such cases a board position is represented by one- or two-dimensional vector in a plain square-by-square manner, without any preprocessing or feature extraction. Each board square is considered as an individual, elementary game feature.

Various game-specific feature sets and game representations were proposed in the literature. Some of them are briefly recalled in this book, particularly in chapter 8 devoted to efficient game representations, and in chapter 6, which presents selected CI achievements in games. Generally speaking, the more elaborate the representation and the corresponding features, the more domain knowledge and human intervention must be involved.

The Form of the Evaluation Function

Typically, the evaluation function is represented as a linear or nonlinear combination of the above-mentioned atomic game features. In the simplest (and also the most popular) case a weighted linear combination of features is considered. More precisely

$$P(s, w) = \begin{cases} MAX \text{ if } s \text{ is a win} \\ MIN \text{ if } s \text{ is a loss} \\ 0 \text{ if } s \text{ is a draw} \\ V(s, w) \text{ for all other states } s, \end{cases} \tag{7.1}$$

where

$$V(s, w) = \sum_{i=1}^{n} w_i x_i, \tag{7.2}$$

and values of MIN, MAX are chosen, so as to fulfil the following conditions:

$$\forall s \in NT \ \forall w \quad MIN < V(s, w) < MAX, \quad \text{and} \quad MAX = -MIN \tag{7.3}$$

In the above, NT denotes the set of nonterminal states in the game tree, $V(s, w)$ is the value of position s calculated with real weight vector $w = [w_1, \ldots, w_n]$ and $x = [x_1, \ldots, x_n]$ is a game state feature vector created by mapping a game state into a number of integer values. Such a mapping, may for example, be equal to 1 if a given feature exists (a certain condition is fulfilled) or 0, otherwise. In the case of numerical features an integer value representing this feature (e.g. the number of pawns, the number of kings, the number of possible moves, etc.) is considered. Assuming that the set of features $\{x_1, \ldots, x_n\}$ is complete and optimized, in a given game situation a particular subset of active features among $\{x_1, \ldots, x_n\}$ is defined by assigning nonzero weights to these features.

For practical reasons, $V(s, w)$ in (7.2) may be additionally "squashed" by a hyperbolic tangent function. One of the possible implementations is the following [242]:

$$V(s, w) = a \cdot \tanh \left(b \cdot \sum_{i=1}^{n} w_i x_i \right), \qquad a = 99,\ b = 0.027 \qquad (7.4)$$

with $MIN = -100$ and $MAX = 100$.

Parameter a equals 99 to guarantee that $V(s, w) \in (-99; 99)$ and b equals 0.027 to prevent $V(s, w)$ from growing too fast and to cause $|V(s, w)|$ to approach 99 only when the sum $\sum_{i=1}^{n} w_i x_i$ in equation (7.4) lies outside the interval $(-99, 99)$. Note that although $V(s, w)$ is nonlinear, the core of this function is a simple weighted sum of game state features, and $\tanh(\cdot)$ is used only for technical reasons in order to limit its range to $(MIN + 1, MAX - 1)$ interval.

The evaluation function of the form (7.4) can be implemented by a neural network architecture with n input units (one unit per each feature) and one output unit, without hidden neurons. In the output neuron the hyperbolic tangent is applied as internal transformation between input and output signal. Values of (7.4) in nonterminal states depend on the weight vector w, which is subject to change in the learning process. In other words the learning task consists in optimizing the weights in the linear combination of atomic features.

An evaluation function can also be represented in a nonlinear form, in particular as an MLP with one or two hidden layers. Although in theory using nonlinear representation has an edge over the linear one, in practice this advantage is not easy to achieve. On the contrary, using an MLP with hidden layers may cause several inefficiencies in the learning process. First of all, due to high network's complexity, time requirements of the training process may be very demanding or even prohibitive. Second of all, for some methods (e.g. TD-type ones) problems with convergence come into play. Both these issues are further elaborated on in chapter 9 related to efficient TD training schemes.

Taking into account the above limitations, the linear form (7.2) or (7.4) of the evaluation function is more frequently encountered in the literature than nonlinear representations. One possible realization of the linear evaluation function is the *weighted piece counter* (WPC), formulated in the following way:

$$V(s, w) = \sum_{i=1}^{board\_size} w_i y_i^s, \tag{7.5}$$

where $y_i^s, i = 1, \ldots, board\_size$ denotes the value of square i in state s. In the simplest case, when the pieces are homogeneous, as is the case in Go and Othello, for example, y_i^s is typically assigned a value of $+1$ if square i is occupied by a player's piece, -1 if an opponent's piece is located on square i, and 0, otherwise. In such a case the WPC can be regarded as "almost" knowledge-free position evaluator, where the word "almost" stems from the fact that WPC is built on heuristic domain knowledge about the importance of the piece count difference. Weights w_i in (7.5) can be constant through the entire game or vary in time depending on the game advancement or player's game situation. In the case there is more than one type of pieces used in a game, the representation becomes less and less domain-independent since the values of the board squares reflect the relative strengths of the pieces. For instance, in checkers $y_i^s \in \{-K, -1, 0, +1, +K\}$, where $\pm K(K > 1)$ denotes king's value, and is based on the heuristic knowledge that the king is more important than a simple checker (e.g. $K = 1.5$ or $K = 2.0$). Often the WPC representation takes advantage of game's symmetries, by forcing the same weight for the isomorphic board squares, e.g. the four corners of Go or Othello boards. Applying these additional symmetry constraints allows lowering the number of weights to be developed in the learning process and makes board evaluation invariant to rotations and reflections.

The Learning Method and Its Parameters

Learning the set of optimal coefficients $w_1, \ldots, w_n$ of linear or nonlinear evaluation function can be performed in various ways. The most popular approaches include neural network backpropagation training (for truly nonlinear representation) or the delta-rule training (in the case of one-layer Perceptron), or the use of TD methods for network's weights update. Another possibility is to use coevolutionary-based optimization, in particular the local hill-climbing algorithm. These methods, briefly described in chapter 5, are frequently referred to in this book. Other popular choices are the least-mean square fitting and other local or global optimization techniques.

The success or failure of using the above methods usually depend on the implementation details, such as the size of the hidden layer in neural representation, form of the evolutionary operators, or the values of several internal methods' parameters (learning coefficient, momentum rate, population size, mutation probability, etc.). Generally speaking, the rules of how to effectively apply CI methods to evaluation function learning are similar to those related to other application domains, in which an unknown real-valued function is to be approximated through the learning process.

There is, however, one thing worth particular emphasis in the context of evaluation function learning: the initialization of weights in the training process. Regardless of the applied method (being either TD, neural nets or evolutionary approach) the choice of the starting point is in most cases crucial for the final outcome of the learning process or at least can significantly speed up training. Usually these initial settings are based on human expert knowledge. Several authors stressed the key role of the educated guess when defining the starting point for the training. For example Hughes [156, 157] underlined the significance of the nonzero initialization of the piece difference weight in his Brunette24 checkers player (chapter 6.2). When this gene was set to zero, the evolution could not start, due to a very poor gradient. Also Baxter, Tridgell and Weaver [18, 19] pointed out the importance of proper system's initialization in KnightCap's success (section 6.3.1).

Summary

A discussion on evaluation function design presented in this chapter is determined by the assumption that either a predefined set of presumably relevant game features (usually painstakingly designed by human expert players) is used or a raw board representation (usually in a square-by-square manner) is applied. In both cases the key issue is the choice of the weight coefficients and the form (linear vs. nonlinear) of the evaluation function. In a typical AI approach both aspects are left for human intervention and are usually designed by hand or in an arbitrary manner. The CI methods, on the other hand, rely on learning with little or no human support whatsoever.

Even though the problem of evaluation function learning is quite well researched, there is still some room for improvement and potential breakthroughs. One of the especially tempting possibilities is application of learning methods, which focus not only on the weight assignment, but also on the construction of a suitable form of the evaluation function, optimally fitted to the game representation used.

Furthermore, one may take a radical approach and try to autonomously develop such a meaningful value function starting completely from scratch through defining the self-directing and game-independent process that would lead to the construction of a close-to-optimal set of board (game) features. This issue was discussed by Paul Utgoff who stated in [329]: "Constructing good features is a major development bottleneck in building a system that will make high quality decisions. We must continue to study how to enhance our ability to automate this process." Utgoff suggested that game features should be overlapping and form a layered, hierarchical system in which more complex features are built based on simpler ones.

Examples of the systems capable of autonomous or semi-autonomous learning with limited initial knowledge are presented in chapters 13 and 14, which continue and extend this chapter's considerations.

8

Game Representation

One of the fundamental topics in intelligent game playing is the problem of efficient game representation. In board games a square-by-square description of the board in the form of a linear vector is often applied. Usually the value of each element defines the presence of a (particular) piece and a sign denotes the side to which the piece belongs, with a value of zero representing an empty square. Such an approach, although well suited for computer representation and feasible from the computational point of view, is not necessarily the optimal one in terms of capturing specific game features. These features are often reflected in the form of 2D patterns, rather than linear vectors. Intuitively, when talking about efficient game representations, the ones accounting for spatial, two-dimensional relations should also deserve attention.

Certainly, using 2D representation that takes into account the context (neighborhood) of a given square contradicts, in some sense, the attitude of knowledge-free learning promoted in this book. However, in the author's opinion, unless the proposed representation is tailored to particular game, such a deviation from the pure knowledge-free attitude is still relevant for the scope of the book, since it fits the autonomous, game-independent learning paradigm. In other words, if representational enhancements are rather general (e.g. using 2D instead of 1D representation or applying board symmetries to simplify the learning process) they may be regarded as straightforward and *universal* extensions of raw game representations.

In the case of card games, basic representations include each card's value (*two, three, ..., Ace*) and suit (♠, ♡, ♢, ♣) as well as the assignment of cards into particular hands and into public or hidden subsets, depending on the game rules. In the course of learning, besides acquiring this basic information several other more sophisticated game features need to be developed by the learning system. Typical examples include the relevance of groups of cards of the same value or sequences of cards within one suit. Suitable representation allows faster and easier discovery of these underlying concepts.

Some of the game related representational issues, concerning both board and card games, are discussed in the remainder of this chapter in the context of particular games. This topic seems to be especially important in neural

J. Mańdziuk: Knowledge-Free and Learning-Based Methods, SCI 276, pp. 99–119.
springerlink.com   © Springer-Verlag Berlin Heidelberg 2010

network example-based learning, in which some of the games' peculiarities can be enhanced by suitable game representation.

The most common representational issues include locality of particular games (discussed in chapter 8.1) and symmetries of the board (chapter 8.2). Other approaches reflect specificity of games phases (opening, mid-game, endgame) or relations between various game features (e.g. their concurrent occurrence). Another issue concerns the relation between position representation and the learning method used. Generally speaking, particular learning methods require appropriate game representations well suited to the implemented learning paradigm. An example of an approach tailored to a specific method, within the incremental learning framework is presented in chapter 8.3.

A discussion on games' representations is concluded by a case study related to the influence of deal representation on the effectiveness of neural network approach to solving the Double Dummy Bridge Problem.

Effective game representation is one of the decisive factors in successful design of game playing systems and there is quite a lot of room for research aimed at direct comparison of various game representations while keeping other aspects (training method, evaluation function, ...) fixed within agreeable standards. On the other hand, in the spirit of a knowledge-free attitude to game playing advocated in this book, special care should be devoted to generality of proposed solutions, since the more specific the representation, the possibly less universal and more game-oriented the implemented system.

8.1   Local, Neighborhood-Based Representations

One of the well-known examples of local, neighborhood-based representation is Chellapilla and Fogel's Anaconda aka Blondie24, analyzed in chapter 6.2. In this experiment the entire checkers board was divided into all possible (highly overlapping) squares of sizes $3 \times 3, 4 \times 4, \ldots, 8 \times 8$ and each of these squares was passed separately to the designated first hidden layer neuron. As it was mentioned in chapter 6.2 one of the underlying messages from Blondie24 experiment was the crucial role of this sparse connection pattern between the input layer and the first hidden layer. Initial attempts to coevolve a high profile checkers player using typical fully-connected two-hidden-layer MLP without additional, selectively connected hidden layer presented in [63], were less successful than the local board representation finally used.

Another interesting study related to different ways of board representation was recently proposed by Mayer [220] in Go, where three game codings of 5×5 board were compared. Since Go is solved for this size of the game [331] it was easy to analyze the effects of training obtained for all three tested representations not only by playing against test programs, but also by comparison with the infallible, perfect player. Neural networks representing Go players were trained in a self-play mode with TD(0) algorithm modified so as to calculate the weight changes with respect to previous move of a given player, i.e. a comparison was made with the move executed two ply earlier:

$$w_{t+2} = w_t + \eta[\gamma V_{t+2} - V_t]\nabla_{w_t} V_t \qquad (8.1)$$

with γ set to 1.0 in all experiments.

The other important modification to typical TD-learning approach was using two different rewards: one for the winning side (equal to +1) and the other one for the losing side (equal to 0). This way, not only the winning side (say black) learned that it played well, but also the opponent (white) received a signal of playing poorly. Since the self-training network played both sides, the above twofold reinforcement (different for each color) prevented the possible overestimation of black's strength in case a favorable result for black was caused mainly by poor play of white.

Mayer stated that typical implementation of TD-learning following [319] yielded weaker results compared to the representation that included proposed modifications. Further exploration of this issue, also in other games, seems to be an interesting research topic.

The major goal of Mayer's work was comparison of the effectiveness of three methods of coding Go board named *Koten, Roban* and *Katatsugi* - see fig. 8.1 for graphical description.

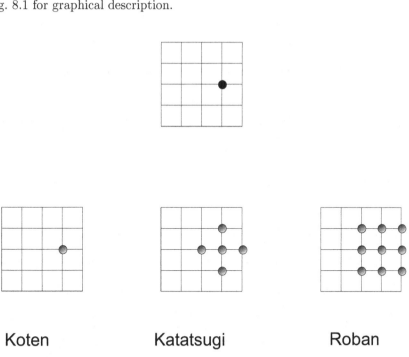

Koten   **Katatsugi**   **Roban**

Fig. 8.1. Local board representations in Go [220]. The bottom figure presents three representations of the field marked black in the top figure: *Koten* - the leftmost picture, *Katatsugi* - the one in the center, and *Roban* - the rightmost one. See description within the text for details

Koten is the simplest, traditional coding in which a player's stone is coded as 1, opponent's stone as -1, and 0 denotes an empty field. The input vector is composed of 25 elements, one per each field. Certainly, no explicit information about the neighborhood or other topological properties of the board are taken into account in this coding.

In *Roban* representation, the 3×3 neighborhoods are considered for all possible locations in which such neighborhood exists in non-degenerative form (there are 9 such fields). A value representing the center field is calculated as a mean value of all its neighbors and field's own value, i.e. the mean is calculated over nine fields as depicted in fig. 8.1. In other words, each field contributes with the value $\frac{1}{9}$, $-\frac{1}{9}$, or 0, respectively for the player's stone, the opponent's one, and an empty field. These 9 values calculated in the above described way are fed into the neural network. Additionally, the average over the complete 5×5 board is calculated as well, and presented to the network as the final 10*th* input.

In *Katatsugi* representation, a value of each field is calculated as a weighted average over its four neighbors and its own value, with all the weights equal to $\frac{1}{5}$. Note that with the fixed weights the range of possible values for fields lying at the border is smaller than for the inner fields, which potentially allows the network to distinguish between these two cases [220]. The *Katatsugi* input vector is composed of 25 elements.

For each of the three above described representations neural networks were trained by TD(0) method in a self-play mode with either ε-greedy policy or *softmax-creativity* method (described in section 9.1.4) with adequately chosen architectures (MPLs of the size $25 - 25 - 1$ in the case of *Koten* and *Katatsugi*, and $10 - 20 - 20 - 1$ for the *Roban* representation). Each network was independently trained 20 times, starting from randomly chosen weights. The effectiveness of training was sampled every 50 000 games and confronted with three external programs: a random player, a naive player - of strength compared to a novice human player, and GOjen player - of strength compared to an amateur human.

In the self-playing mode with ε-greedy policy the two 25-input-based representations were comparable to each other (in terms of attained level of play after one million training games) and they both clearly outperformed the *Roban* coding. The situation was slightly different when the creative choice mode was considered. Here the *Katatsugi* coding took the lead, followed by *Koten*, and finally by *Roban* representation. Interestingly, the above order was not confirmed in the round-robin tournament between these 6 heuristics (three codings times two types of learning schemes), where the absolute winner was *Koten* with softmax-creativity, followed by *Koten* with ε-greedy and two *Katatsugi* and two *Roban* networks (with softmax-creativity beating ε-greedy in all three representations).

In comparison with external programs the best creative *Katatsugi* networks at the search level of depth 1 achieved a 98% score against random player, and 100% at depth search of 2. These networks also convincingly conquered the GOjen player, with the winning rate of 65%. Moreover, they managed to learn how to play optimally the initial move, which is at the center of the board.

The results indicate that board representations in which for each field an explicit part of the representation, related to that field, is constructed (either as a particular field's value (*Koten*), or a combination of this value and values within some neighborhood (*Katatsugi*)) dominate over the more compact *Roban* representation. Although the results are convincing for 5×5 board, in order to be conclusive they need to be confirmed on greater boards.

Additionally, some modifications to the two nonstandard codings can be proposed. For instance, in *Katatsugi* one can think of using different weights for border and corner fields; in both *Katatsugi* and *Roban* it seems reasonable to consider using higher weight for the field under consideration compared to its neighbors' weights. Despite possible modifications which might be taken into account when continuing this work, this piece of research is definitely an interesting attempt to verify the properties and efficacy of alternative Go representations.

8.2 Invariants-Based Representations

Two of the games considered in this book, namely Othello and Go, are invariant to several symmetries - reflections and rotations of a game board. In both games a position can be reflected symmetrically along horizontal, vertical or any of the two main diagonal axes as well as rotated by the $k\frac{\varPi}{2}, k = 1, 2, 3$ degrees without affecting the board evaluation. Taking into account these invariance properties can simplify the form of the evaluation function (by lowering the number of weights) and also help keeping it consistent (by applying the same weights/assessments to isomorphic game positions).

In particular, in the case when the evaluation function is represented by a neural network the idea of weight sharing leads to significant simplification of the network's training process by lowering the number of (independent) weights that need to be properly set. The idea of weight sharing was initially proposed in the pattern recognition domain, in the context of neural systems prone to image rotations and translations [118, 180, 181]. In game domain weight sharing became popular in the mid 1990s, when several successful examples of this approach have originated, especially in Go [293] and Othello [183]. Recently a resurgence of interest in this field can be observed as the number of papers that utilize this technique increases. Some recent examples of game codings that benefit from existence of board symmetries in Othello and Go are presented below.

Othello

In the game of Othello, board representation method can capitalize on the existence of 4 symmetry lines, which form 8-way symmetry. This allows treating particular squares as indistinguishable, since from the algorithmic point of view they bear the same meaning. For example squares $b3$, $b6$, $c2$, $c7$, $f2$, $f7$, $g3$, and $g6$ form such an "equivalence class" in the board squares space. Consequently, coefficients assigned to these fields in the WPC evaluation function should be pairwise equal and on the algorithmic level all of them may be treated as one parameter. Similarly, any game pattern covering one of these fields can be automatically transformed into other parts of the board by using symmetry lines.

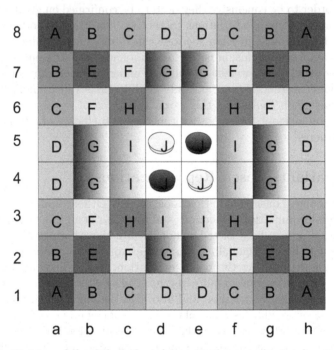

Fig. 8.2. Division of the Othello board into 10 abstract classes, denoted from A to J, according to board symmetries. In Othello-related literature elements of class B are usually referred to as *C-squares* and the E-class ones as *X-squares*

One of the recent approaches to Othello, which benefits from symmetries of the board is proposed in [240], where board squares are grouped into 10 classes based on the 8-way symmetry properties as depicted in fig. 8.2. Consequently, the evaluation function is composed of 15 primary elements only, among which 10 are used for a raw board representation - one element per class, and the remaining 5 are: the number of possible player's moves, the number of possible opponent's moves, the number of discs that cannot be flipped by the opponent, the number of player's discs on the board, and an indication of which side is to move. The authors propose using three instances of the weight vector corresponding to the opening, mid-game, and endgame phases, respectively. The opening stage is defined as lasting from initial position until a disc (of any color) is placed at the edge of the board. The mid-game stage ends when two stones of the same color are present in the corners. This marks the beginning of the endgame phase.

Since the main goal of [240] is an introduction of Monte Carlo modification to TD(λ) algorithm, it is not possible to thoroughly judge the efficacy of the proposed board representation. A direct comparison with the WPC evaluation function relying on all 64 squares as independent features is one of the prospective goals. Another possibility of comparison is using each square as an independent feature during the training and once it is completed setting all weights assigned to squares within the respective class to one common value being either the average, minimum, maximum, or median value.

Symmetry-based weight sharing in Othello has recently been also utilized in the self-play TD-learning framework [215]. In this approach the majority of connections between input and hidden units are shared in the network, taking advantage of the 4-way symmetry. For any board position there exist three other equivalent positions to be considered: two of them obtained via reflections relative to each of the two main diagonals and the remaining one by using a superposition of both the above reflections (in any order). The input layer is composed of 64 neurons and each of them is connected to four hidden layer neurons through shared weights. Each of the hidden layer units represents one of the four equivalent reflections. A common set of 64 shared weights incoming to each hidden neuron is renumbered according to the symmetry properties represented by this neuron. The bias weights incoming to hidden units and the hidden-to-output connections are also shared by the four hidden neurons, which results in only two, instead of eight, additional weights. The last network connection is the bias weight coming to the output unit.

The network's weights were trained by self-play TD-learning for 250 000 games with a relatively low learning rate of 0.0001. The range of weight initialization was also unusually narrow. The weights were chosen randomly from the interval $(-10^{-4}, +10^{-4})$. The ε-greedy policy was adopted with $\varepsilon = 0.1$. Despite using as little as 67 unique weights only, the network was able to perform as well as the CEC 2006 Othello Competition [198] champion, implemented as an MLP with 32 hidden nodes and 2113 weights.

Go

Similarly to Othello, the game of Go exhibits four symmetry lines, as well (one vertical, one horizontal, and two diagonal). Hence, the board can be rotated or reflected in 8 different ways, without affecting assessment of the position. Moreover, Go possesses the quasi-shift-invariance property, since it is the local arrangement of stones rather than their absolute location, that usually causes the placement of the next stone on a certain field in the neighborhood of the stone formation. This feature is attributed to strong locality of Go and makes a significant difference between this game and Othello.

Board symmetries were taken into account in several approaches to Go, e.g. by Schraudolph et al. [293, 295], where representation invariant to color reversal, reflection and rotation of the board was used. The authors applied a specially designed neural network, trained with TD(0) method with the local reinforcement signal assigned to each individual intersection on the board. As part of the neural architecture, a fixed-size receptive field with weight kernel was used to scan the board representation and make the system translation invariant. The input layer was composed of 82 neurons (one per each intersection point in 9×9 Go plus additional bias). Each board-representing input unit received a signal of ± 1 depending on which side occupied the respective point, or 0 in the case of empty point. The output layer was composed of 81 neurons and, if the stone had been captured on the respective intersection, a reward of ± 1 was received during the game. Likewise a reward of ± 1 was assigned to each output unit after the game had been completed, depending on to whom (white or black) the territory ultimately belonged to. This way the network predicted the final state of each point on the board rather than the overall game score.

Another possibilities of specific board representation in Go were proposed by Enzenberger in *NeuroGo ver. 2* (NeuroGo2) [90] and *NeuroGo ver. 3* (NeuroGo3) [91] programs. Both systems took advantage of domain knowledge in the form of specialized modules. In the former system a feed-forward neural network was trained by self-play TD(0) method with weights being changed through backpropagation method. In 15% of the training moves the choice was made according to the Gibbs distribution (9.5), introduced in section 9.1.4, with $T = 0.35$. Go position was transformed into a set of strings (chains of stones) and empty intersections. Strings as well as empty points were treated as single units. In the input and in the hidden layer an arbitrary number of neurons was assigned to each unit. Generally speaking, the more neurons assigned to individual unit the more precise and richer the description of mutual relations between the units.

The *a priori* knowledge was delivered to the system by three expert modules: relation expert, feature expert and external expert. The relation expert built a connectivity graph between units in a way which was invariant to translation and rotation of the board. The feature expert detected predefined features of single units and changed the activations of the corresponding units accordingly. The external expert used its own evaluation of the board (not bound with the training process) and had the ability to override the network's output decisions.

In the output layer sigmoidal neurons were used (one per each board intersection) to predict the probability that the respective point would belong to black at the end of the game. NeuroGo2 was tested on 9×9 board against *Many Faces of Go* (MFoG) level 8, playing on equal terms after about 5 000 training games, with 24 neurons per unit.

The next version of NeuroGo [91] was also relying on self-play TD(0) learning combined with backpropagation. Similarly to previous version of the system, in the training phase, in 15% of cases the move selection procedure followed the Gibbs distribution, but this time with $T = 4.0$. The architecture was significantly extended by implementing a soft segmentation mechanism, which allowed division of the game position into parts covering presumably independent subgames. The method was more flexible than using a fixed-size convolution mask with shared weights (as was the case in [293]), or using dynamic receptive fields of adaptable size fitting the size of strings of chains (units), applied in NeuroGo2. In particular, the proposed soft segmentation method was capable of managing the distance relationships among a group of loosely connected chains of stones, which might become one coherent chain later in the game. Appropriate handling of these relations, represented as probabilities of potential connections (denoted as *connectivity maps* in the paper) was of great importance in the overall board assessment.

In effect, with additional input feature preprocessing, the system's playing strength was estimated slightly below the level of 13kyu, which was one of the best computer Go results at that time (2003).

Board invariance properties have recently been also used by Blair [34] who proposed application of Cellular Neural Networks together with a novel weight sharing scheme called *Internal Symmetry Network*. The implementation of weight sharing mechanism was inspired by existence of similar phenomenon in quantum physics. The system was trained in a self-play regime using TD-learning with $\lambda = 0.9$. The overall board assessment P was equal to the sum of individual assessments of all board locations. Moves were chosen according to Boltzmann distribution (9.5) with probability of each move proportional to $e^{\frac{P}{T}}$, with $T = \frac{1}{4}$ during training and $T = \frac{1}{20}$ in the evaluation phase. The system was trained on 9×9 version of Go and subsequently tested on both 9×9 and 19×19 boards. According to the authors the network's performance was quite reasonable and the system was capable of making capture, blocking, and threat moves. On the 19×19 board in majority of the cases the move chosen by a human expert was within the top 10-20 best moves declared by the network.

Another recent examples of symmetry exploitation in Go are Stern et al. [308] and Araki et al. [8] papers discussed in chapter 10. In both approaches characteristic patterns which may be transformed into each other by rotation, reflection, color reversal, or any combination of the above, are treated as the same patterns.

8.3 Incremental Learning

The idea of *incremental learning*, in its general formulation, consists in the ability of the trained system to acquire new information without destroying already possessed knowledge. Incremental training methods in the example-based supervised neural network learning framework were developed in response to the so-called *catastrophic interference* or *catastrophic forgetting* [117, 223] appearing in sequential learning. This undesirable effect is considered to be one of the greatest impediments in building large, scalable learning systems based on neural networks.

The idea of incremental training can be realized in various ways, depending mainly on the specificity of the learning task. One of the possible approaches can be described in the following way. The learning starts off in an environment smaller than the target one. During the training process the environment is gradually increased - up to desired size - and after each change of size some number of training examples, specific to this increased environment, is added to the training set, in order to capture new features that arose in this larger, more complicated environment. The claim is that after some training time, the system should be able to recognize features, which are *invariant to the size (degree of complication) of the environment*. In such case these size-invariant features will be shared among several instances of the environment and used during the training process to make generalizations about the learning task, hence preventing catastrophic forgetting.

Obviously, the above idea cannot be applied immediately to every learning task. Even with the application domain restricted to mind board games, not all of them are well suited for this paradigm. Playing chess, for example, on a 6×6 board sounds odd, to say the least. On the other hand, there are games that are well suited to the above described incremental training mechanism. One can easily redefine checkers, Othello or Go to be played on smaller or greater boards than their original specification.

For the games of this type, with no pieces differential, the idea of incremental training can be implemented in a quite straightforward way. Consider a game which is played on a board of size n. The training process begins on a game board of smaller size $k, k < n$ and once the agent learns how to play or solve problems on this board, the board size is increased to $k + t$, where t depends on particular game ($t \in \{1, 2\}$ for majority of popular games). Then the agent is retrained *in a limited manner* based on the extended set of problems presented on the increased board. The retraining procedure is significantly shorter than the regular training performed on board of size k. Then a board size is increased again and the agent is retrained, and so on. The whole procedure is stopped when retraining on board of size n is completed.

The underlying idea of the above scheme is that after some preliminary phase learning should become relatively easier and solutions to problems defined on larger boards should be developed by the system based on already defined solutions to problems stated on smaller-size boards. Hence, *subsequent learning would mostly involve efficient use of previously acquired knowledge*.

The above described training scheme was applied to Othello on 6×6, 8×8 and 10×10 boards (the board of size 10×10 was the target one) [203, 204]. Representation of board positions was significantly simplified compared to typical n^2-input-based one. Instead of presenting the entire board only four n-element vectors were used to represent a given game position (see fig. 8.3).

Each training example was composed of a game position and its evaluation after a certain move X was executed. The four above-mentioned vectors were used to represent data corresponding to the row, column and two diagonals that included the field X. For example in figure 8.3(a) the 8×8-board position with the black move at $c7$ is represented as four vectors denoted respectively by R (row), C (column), D_1, D_2 (two diagonals). Observe that each vector is extended to the length of $n = 10$, which is the target size of the board, by adding unoccupied fields. Similarly, in figure 8.3(b) the 6×6-board with the white move at $a5$ is represented in the form of four n-element vectors.

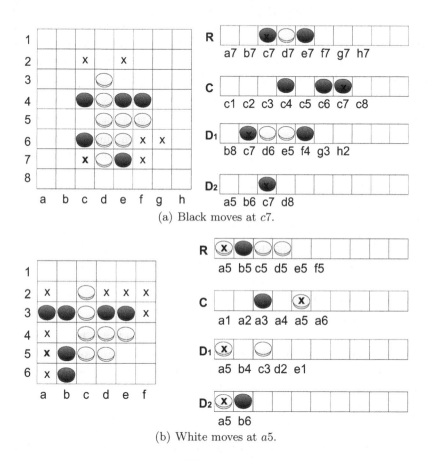

(a) Black moves at $c7$.

(b) White moves at $a5$.

Fig. 8.3. Two examples of a simplified position representation proposed in [203, 204]. Instead of the complete board representation (left parts of the figures), four 10-element vectors are used (right parts of the figures). R, C, D_1 and D_2 denote row, column and two diagonals passing through the field where the move is to be made. x marks denote legal moves, respectively for black (top figure) and white (bottom figure). Bold x marks denote moves considered to be executed in the respective positions (these moves were selected at random among all legal moves). See description within the text for more details

The MLP network was trained with backpropagation method on examples extracted from 6×6 Othello positions and subsequently retrained, in a limited manner, on extended training set containing 8×8 and - in the final step - 10×10 positions. The training goal was to point out the best move in a given position. The results were compared with a full backpropagation training carried out on 10×10 board examples (also represented as four 10-element vectors). The amount of time required for incremental training was considerably lower than in the case of the full training. Also numerical results showed a slight improvement over a one-shot training procedure: after incremental training the network responded with the first, the second, and the third best moves (selected according to applied heuristic) respectively in 40%, 4% and 15% of the cases, whereas for the complete backpropagation training on 10×10 board the respective figures were equal to 34%, 19% and 17% [204].

The advantage of proposed representation of a game position over the complete board representation is both simplicity and size-independence (given the ultimate board size n), which allows time-efficient implementation of incremental training paradigm introduced above, and potentially leads to size-independent generalization of game features.

On the other hand, the side effect of great simplification of position representation is that different game situations may have common representation, since the mapping between positions and vector-based representations is not bijective. In order to alleviate this indistinguishability problem, the above-proposed representation may be extended by adding other vital, non-global features, for instance, the two vectors representing the main diagonals.

Two Recent Approaches

A more recent example of knowledge transfer between two instances of the same game was presented by Bahçeci and Miikkulainen [14]. The task chosen for the experiment is a simple one-player game consisting in placing a fixed number of white stones on a square game board so that as many of them as possible attack the black stone which is randomly placed on one of the board fields at the beginning of the game. Attacking black stone is equivalent to placing a white stone in the same row or in the same column as the black stone, within a predefined distance from it.

A simplified version of the game is played on a 4×4 board with no distance limit, whereas the target game instance is a 12×12 version with attacking distance less or equal 3 in both rows and columns. Game patterns (features) in a simplified game version are constructed as *compositional pattern producing networks (CPPNs)* [306], which are represented as one-layer neural networks with three inputs (coordinates x and y of the square under consideration and additional bias input fixed to 1) and one output (representing a value of the pattern at square (x, y)) with either sigmoidal or Gaussian transfer function. Coordinates (x, y) are calculated relative to the center of a pattern, which can be located at any board's square.

All patterns are of size 7×7, i.e. fully cover the game board regardless of the choice of their center square location. Consequently x and y can each take integer values from the interval $[-3, 3]$. A population of 200 such CCPN's was evolved for 45 generations according to the NEAT coevolutionary scheme [307] (described in chapter 14), with specially defined evolutionary operators. Individuals in the initial population were minimally connected, i.e. only one out of three inputs was connected to the output unit. The finally evolved population was then used as the initial population for another coevolutionary process applied to solving the target problem. Evolution in this case lasted for 1 000 generations and the experiment was repeated 20 times and confronted with the results of 20 coevolutionary simulations starting from random populations.

The results confirmed that the population evolved for the simplified problem is a better staring point for the target evolution than a randomly selected population. In particular the transfer-based approach requires fewer generations (totally in both evolutions) to evolve a perfect player.

One of the weak points of the method is manual choice of the simplified version of the game. It would be desirable to automate this process, especially in the context of potential implementation of the algorithm in the General Game Playing framework (discussed in chapter 14). Moreover, application to other, more demanding games is expected to further confirm the value of the proposed ideas.

Another form of incremental learning is the ability to generalize from small game instances (played on small boards on which training is both faster and less expensive) to larger boards *with no requirement for supplementary training.* In such a case learning is independent of a board size and generalization is usually assured by adequate system's design.

One of the promising approaches of this kind has been recently proposed by Schaul and Schmidhuber [290, 291]. The authors applied scalable Multi-dimensional Recurrent Neural Networks (MDRNNs) [142] to learn three simple games played on a Go board: *atari-go* (a simplified version of Go), *go-moku* (also known as *five-in-a-row*), and *pente* (a moderately complicated version of go-moku).

The MDRNNs were trained either by evolution or coevolution proving good scalability in both cases. In particular, the networks were capable to play each of the three games on 19×19 board after being trained on 7×7 board. Promising experimental results were obtained despite learning from scratch, based on raw board representation, and with no domain knowledge involved.

In most of the conducted experiments the test opponent was either a random player or a naive player. The latter one was searching 1-ply, and chose an immediately winning move, if available, discarded immediately losing moves, and picked a move at random, among the remaining ones, otherwise.

Quite surprisingly, the intuition that the smaller the difference between the target and initial board sizes, the better the effects of knowledge transfer between these two game instances, appeared not to be generally true. A correlation between the results accomplished on 5×5 or 7×7 boards and those obtained on larger boards was not monotonically decreasing with the increase of the target board size.

One of the prospects of possible development of the above work is cascading two or more MDRNNs on top of each other in order to allow capturing more complex game features and game properties than in the one network case.

8.4   Double Dummy Bridge Problem

In bridge, special focus in game representation is on the fact that players cooperate in pairs, thus sharing potentials of their hands. In particular, in neural network example-based training the link between players in a pair needs to be provided. In the knowledge-free methods advocated in this book such a link should, ideally, be defined in an implicit way and discovered by the network itself during the training process.

In order to illustrate the impact of various hand representations in the game of bridge let us consider the *Double Dummy Bridge Problem* (DDBP). Recall from chapter 3.3 that DDBP consists in answering the question about the number of tricks to be taken by a given pair of players on condition that all four hands are revealed and all sides play in an optimal way. The problem, at first sight, may not seem demanding (since all cards are revealed), but it apparently is, even for top bridge players.

In [211, 235] the results accomplished by neural networks were compared with the efficacy of professional bridge players in solving DDBP under restrictive time constraints, considering two variants of the problem - the classical one (all four hands are revealed) and a more difficult, though more "realistic" version: two hands of the pair being scored are revealed and the remaining two are hidden - only the *sum* of the opponents' hands is available as a result of subtracting both known hands from the whole deal. Humans visibly outperformed neural networks in the classical variant with *notrump* contracts, but in the remaining three cases (the classical variant with *suit* contracts, or partly covered variant with *notrump* or *suit* contracts) neural networks appeared to be very competitive to humans - see [235] for details.

Despite the above encouraging results the DDBP, due to numerous nuances of bridge playing, is a challenging task for example-based, knowledge-free learning methods. Previous attempts to solve the DDBP with the use of neural nets, based exclusively on raw data [123, 124] were unsuccessful. The authors of the above cited papers stated that "Neural networks in their purest form take the raw input data, and learn to construct appropriate outputs without doing anything more than recalculating the weights of the connections between their nodes. However, it is in practice considerably more efficient to perform a certain amount of preprocessing on the input, so as to construct values representing features which humans consider important in the domain;" These pre-computed features concerned, for example, "specific high-cards or total suit-lengths."

On the contrary to the above statement the results presented in [235] and earlier works by Mossakowski and Mańdziuk [209, 210, 233, 234] strongly support the claim that with appropriate choice of neural architecture and in particular the input representation of a deal, it is possible to achieve a high score in solving the DDBP with no need for any domain-related preprocessing of the raw data. The way a deal was represented in the input layer in the above experiment turned out to be crucial for the quality of achieved results. The following four ways of coding a deal were invented and tested.

26x4 Representation

In the first way of deal representation, 104 input values were used, grouped in 52 pairs. Each pair represented one card. The first value in a pair determined the rank of a card (A, K, Q, etc.) and the second one represented the suit of a card (♠, ♡, ◇ or ♣). Hence, 26 input neurons (13 pairs) were necessary to fully describe the content of one hand - see fig. 8.4.

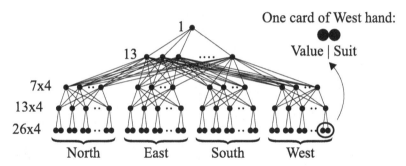

Fig. 8.4. Example neural network architecture with 26x4 input coding. Neurons in the first two hidden layers are connected selectively. They are responsible for collecting information about individual hands

A few schemes of transforming card's rank and suit into real numbers suitable as input values for the network were tested. Finally the rank of a card was represented using a uniform linear transformation to the range $[0.1, 0.9]$, with the biggest values for *Aces* (0.9), *Kings* (0.83) and the smallest for *three spots* (0.17) and *two spots* (0.1). Some other ranges, e.g. $[0, 1]$ or $[0.2, 0.8]$ were also tested, but no significant difference in results was noticed. A suit of a card was also coded as a real number, usually by the following mapping: 0.3 for ♠, 0.5 for ♡, 0.7 for ◇, and 0.9 for ♣.

In order to allow the network to gather full information about cards' distribution, special groups of neurons were created in subsequent layers. For example, the $(26x4) - (13x4) - (7x4) - 13 - 1$ network was composed of five layers of neurons arranged in a way depicted in fig. 8.4. The first hidden layer neurons were focused on collecting information about individual cards. Four groups of neurons in the second hidden layer gathered information about the respective hands. The last hidden layer combined the whole information about a deal. This layer was connected to a single output neuron, whose output range ([0.1, 0.9]) was divided into 14 equidistant subintervals representing all possible scores (the numbers of tricks that could be taken by the NS pair of players).

52 Representation

In the second way of coding a deal, a different idea was utilized. Positions of cards in the input layer were fixed, i.e. from the leftmost input neuron to the rightmost one the following cards were represented: 2♠, 3♠, ..., K♠, A♠, 2♡, ..., A♡, 2◇, ..., A◇, 2♣, ..., A♣ (see fig. 8.5). This way each of the 52 input neurons was assigned to a particular card from a deck and a value presented to this neuron determined the hand to which the respective card (assigned to this input) belonged, i.e. 1.0 for *North*, 0.8 for *South*, −1.0 for *West*, and −0.8 for *East*.

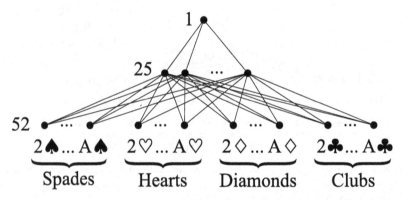

Fig. 8.5. Example neural network architecture with 52 input neurons. In this deal representation standard, fully-connected MLP architectures were applied

Interestingly, it turned out in further experiments that using the same input value (−1.0) for both *West* and *East* hands improved the results. Moreover, hiding the information about exact cards' assignment within the *NS* pair, i.e. using the input value equal to 1.0 for both *North* and *South* hands, yielded another slight improvement.

In this coding there were no dedicated groups of neurons in the hidden layers. Layers were fully connected, e.g. in the $52 - 25 - 1$ network all 52 input neurons where connected to all 25 hidden ones, and all hidden neurons were connected to a single output neuron, whose role was the same as in the previously described case.

104 Representation

The third proposed way of coding a deal was a straightforward extension of the 52 representation to the 104 one. The first 52 input values represented assignments to pairs in a similar way as in the 52 representation, with value 1.0 representing NS cards and -1.0 - cards of WE. The remaining 52 inputs pointed out the exact hand in a pair (value 1.0 for N and W, and -1.0 for S and E). In both groups of input neurons positions of cards were fixed according to the same order (see fig. 8.6).

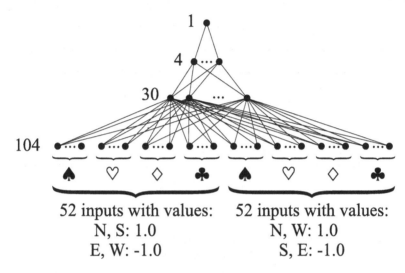

Fig. 8.6. Example neural network architecture with 104 input neurons. In this deal representation standard, fully-connected MLP architectures are applied

Networks using this coding were fully connected and usually contained two layers of hidden neurons, e.g. $104 - 30 - 4 - 1$.

52x4 Representation

The last tested way of coding a deal arose from results obtained by networks that used human estimators of hand's strength (see [211, 235] for the details). These results suggested that it was difficult for networks using the above presented ways of coding a deal to extract information about lengths of suits on particular hands. This information, however, is vital, especially for suit contracts, so there was a need for another representation of a deal, in which lengths of suits on hands would become perceivable by neural networks, although still basing on *raw data*.

In this deal coding 208 input neurons were divided into 4 groups, one group per hand, respectively for N, E, S and W players. Four input neurons (one per hand) were assigned to each card from a deck. The neuron representing a hand to which this card actually belonged received input value equal to 1.0. The other three neurons (representing the remaining hands) were assigned input values equal to 0.0. This way, a hand to which the card was assigned in a deal was explicitly pointed out.

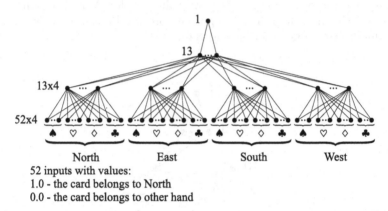

Fig. 8.7. Example neural network architecture with 52x4 input representation. Input and the first hidden layer neurons are not fully connected. The connection pattern allows combining information from redundant and "scattered" input in an efficient way

In this representation each suit on a given hand was represented by 13 input neurons. The number of input values equal to 1.0 among these neurons determined the length of this suit on the hand, so networks using this representation had higher chances to find both long suits and shortnesses (which are very important in bridge), especially voids (no cards in a suit) and singletons (one card in a suit).

There were 4 groups of neurons in the first hidden layer, each of them gathering information from 52 input neurons representing one hand. This data was further compressed in another one or two fully connected hidden layers. All neurons from the last hidden layer were connected to a single output neuron. A sample network using this way of coding a deal, $(52x4) - (13x4) - 13 - 1$, is presented in fig. 8.7. The 3-hidden layer architecture was, for example, realized by the network $(52x4) - (26x4) - 26 - 13 - 1$.

The Trump Suit and the Opening Lead

When talking about deal representation in the input layer, apart from card locations, there are two more issues which need to be addressed. The first one is the way of pointing the network the trump suit, in the case of *suit* contracts (the problem certainly does not exist in the case of *notrump* contracts, which were dealt with by a separate set of networks). This problem was solved by selecting one of the suits (namely the *spades*) as the fixed trump suit and exchanging the cards of the actual trump suit with *spades* cards whenever necessary.

Some limited number of experiments was also performed with trump suits freely chosen, in order to verify the network's ability to adapt to the changes in this selection, but the results were unsatisfactory. In these cases a trump suit for the 52, 104, and (52x4) representations was pointed out by decreasing input values assigned to all cards other than the trump suit (usually by dividing them by 2). The above method could not be applied to the (26x4) representation, because the suit was defined there by a value presented to one of two input neurons representing a card. In that case one additional neuron was added to the input layer and the value presented to it indicated the trump suit. Not surprisingly, the (26x4) networks had problems with perceiving this additional information (only one out of 105 input neurons in total was used to represent this information), and results obtained by the networks of this type trained on deals with various trump suits were worse than the ones obtained by analogous architectures trained only on deals with one, arbitrarily chosen trump suit.

The other important issue is a way of representing a hand making the opening lead. Similarly to the case of the trump suit selection, the problem was avoided by fixing the hand making the opening lead (namely W) in the main experiments. The idea of fixing the opening lead hand was applied in both *suit* and *notrump* experiments.

Since in some deals the number of tricks varies when a hand making the opening lead is changed, some limited number of experiments was performed with the opening leads from the other hand, i.e. E (with adequate change of the hand playing the contract). In order to distinguish the opening lead hand, in the (26x4) representation four additional input neurons were used. Each of them was assigned to one hand and had a nonzero value, if that hand was to make the opening lead. In the remaining representations there was no need for additional input neurons. In these cases the problem was solved by fixing positions of input neurons representing the hand making the opening lead, i.e. each deal was present in the training set twice (e.g. once with the opening lead from the *West* hand and once from the *East*). In the latter case the deal was "rotated" to assure, that always the same input neurons were assigned to a hand that made the opening lead.

Comparison of Representations

In the main experiment all network architectures were trained based on deals from the GIB Library [133] using 100 000 deals for training and (another) 100 000 deals for testing. The best results obtained by the four types of networks for *spades* contracts with fixed opening lead hand are presented in Table 8.1. Each result consist of three numbers $(A \mid B \mid C)$ representing the percentage of test deals for which the prediction error did not exceed 2 tricks (A), 1 trick (B), and 0 tricks (C). The most interesting is certainly the last

number, i.e. the level of perfect answers, however it should be emphasized
that there are deals for which it is very hard (even for experienced human
players) to point out the correct number of tricks. A small difference in the
cards' location, e.g. exchange of two plain cards, can change the result by one
or more tricks. Under these circumstances, the two remaining figures (A and
B) are also worth attention.

This comparison emphasizes superiority of the (52x4) representation. In
the *spades* contracts, the result (99.80 | 95.54 | 50.91) means that the network
answered perfectly in over 50% of deals and was wrong by more than one trick
only in 4.46% of deals and by more than two tricks in 0.2% of deals. Further
improvement has been accomplished by using each training deal twice with
the opening lead respectively from W and E hands. In such a case the rate of
perfect answers increased to 53.11%. The worst results were obtained by the
network using the (26x4) representation. The efficiency of the two remaining
ways of coding was comparable.

For the *notrump* contracts the results of all neural architectures were visibly poorer than those obtained for the *spades* contracts (see Table 8.1). A
detailed analysis of the reasons for such deterioration is one of the future
research goals. One of the hypothesis is that weaker performance arose from
the fact that neural networks used in this experiment were restricted to thorough analysis of cards distribution, which included location of honors and
lengths of suits, but did not include any play phase simulations. Due to a
different specificity of *notrump* and *suit* contracts, the latter ones are "better suited" for simulation-free estimations made by neural networks than the
notrump contracts, which are more sensitive to exact location of particular
cards (especially honors) and more significantly hindered by the absence of
roll-out simulations [86].

Table 8.1. Comparison of the best results obtained with various approaches to
coding a deal for *spades* and *notrump* contracts with the opening lead from the
West hand (the first five rows of results). The last row presents the results for
spades contracts in the case of duplicating each deal in the training set (with
adequate rotation of hands) in order to present both possibilities of the opening
lead hands (W and E)

The network	Results for *notrump* contracts			Results for *spades* contracts		
(26x4)-(13x4)-(7x4)-13-1	93.87	75.70	31.04	97.67	84.24	36.82
52-25-1	96.07	80.88	34.66	98.77	88.00	40.13
104-30-4-1	95.64	79.63	33.74	98.61	87.17	39.21
(52x4)-(13x4)-13-1	97.34	84.31	37.80	99.78	95.00	50.03
(52x4)-(26x4)-26-13-1	96.89	83.64	37.31	99.80	95.54	50.91
(52x4)-(26x4)-26-13-1				99.88	96.48	53.11

The experiment confirmed intuitions concerning high dependency between the input representation and the quality of neural network-based classifications. The detailed discussion on various aspects of obtained classifiers is presented in [235]. Apart from promising numerical results the networks exhibited an interesting representation of bridge-related concepts in their weight spaces. This in turn lead to autonomous discovery of the importance of honors (especially Aces), suit lengths, and other more sophisticated findings, for example the possibility of the *finesse* or the relevance of a trump suit. The above "spontaneous" *knowledge discovery* in the trained networks is discussed in chapter 13.

Efficient TD Training

Several successful applications of TD-learning methods in various games, which include Samuel's checkers player [277, 279], Tesuaro's TD-Gammon [317, 318, 319], Baxter et al.'s KnightCap [18, 19], Schraudolph et al.'s Go program [294, 295], or Schaeffer et al.'s TDL-Chinook [288], to name only the best-known examples, confirmed wide applicability of this type of learning in game domain. Quite a lot of research papers devoted to TD-learning and games were published, in which particular attempts are thoroughly described focusing on various aspects of TD paradigm. Based on the amount of the literature one might suppose that all relevant questions concerning this type of learning have already been answered and if there exist any ambiguities they must refer to secondary issues or specific implementation details.

However, quite surprisingly, the research reports do not agree even on very basic, fundamental issues as, for example, the effectiveness of self-play TD versus training with external opponents. In the former case the basic questions refer to the exploration-vs.-exploitation dilemma, which consists in a tradeoff between playing along already known, optimal (to the knowledge of the learning program) lines and the exploration of new possibilities at the risk of playing moves weaker that the optimal ones.

In the case of training against external opponents the key issues are related to the choice of the trainers: the number of them and their playing strength. Intuitively, the more training opponents the better, since variety of the trainers leads to variety of possible game situations that the learning program is faced with. On the other hand, at some point the problem of convergence of the training process comes into play. As far as the strength of the trainers is concerned, the intuitive approach is to use opponents of comparable or slightly superior playing level than the learning program, gradually increasing their average playing abilities along with the development of the trainee. Another possibility is to use random or pseudo-random opponents[1]

---

[1] The distinction between *random* and *pseudo-random* players will be cleared in section 9.2.6.

J. Mańdziuk: Knowledge-Free and Learning-Based Methods, SCI 276, pp. 121–153.
springerlink.com   © Springer-Verlag Berlin Heidelberg 2010

in the initial phase of training and replace them with stronger players as the training process progresses. Last but not least problem is weighing the pros and cons of training against human players vs. artificial ones.

Finally, the choice of the specific TD-learning method, being either TD(λ) or TDLeaf(λ), as well as the choice of λ decay schedule have significant influence on attained results and in the literature there are different points of view regarding this matter.

This chapter attempts to address these issues basing to a large extent on the author's experience with TD-GAC - a TD-based program trained to play give-away checkers, introduced briefly in section 6.3.3. The TD-GAC experiment aimed, in particular, at comparison between various training strategies including the variants of learning exclusively on lost and drawn games, learning on all games but with omitting weak moves, or learning with focus on opponents stronger than the learning program by playing several games in a row against the same opponent whenever the program kept losing against that opponent.

Certainly the results, ideas and comments presented in this chapter do not pretend to provide complete and definite knowledge in the above-mentioned areas. The goal is rather to attract readers' interest to these topics since the more theoretical and experimental results are available the better understanding and better outcomes of TD-learning applications in games.

9.1   External Opponents vs. Self-playing

One of the fundamental questions that need to be answered when planning a TD-learning experiment is the choice between self-play training mode and learning based on games played against external opponents (independent of the learning system). Despite very basic nature of this issue, the experiments published in the literature are inconclusive with regard to whether it is more profitable to favor self-playing or rather to train with external opponents. Hence, one of the interesting and challenging issues is further investigation and formalization of the strengths and weaknesses of both training approaches.

9.1.1   Self-playing

The first widely-known example of a TD-learning process relying on a self-playing scheme was Arthur L. Samuel's approach to checkers[2] [277, 279]. The program was equipped with *a priori* defined set of expert features potentially relevant to building board evaluation function, and in the course of self-play learning was able to successfully define the meaningful subset of features and their weights in order to form linear evaluation function.

---

[2] Formally speaking, at that time the term TD-learning was not yet invented, but the learning method used by Samuel was very akin to this type of training.

Samuel's work was partly inspired by Claude Shannon's ideas [296], in particular with respect to applying the minimax search method. Samuel used essentially two learning modes: the first one called *rote learning* consisted in memorizing game positions encountered during the tree search together with their heuristic evaluations backed-up by the minimax. This type of learning appeared to be effective in the opening and endgame phases and the overall judgement of the resulting program was that it played on better-than-average novice player level [314].

The other method, called *learning by generalization*, was of the reinforcement learning nature. In this training mode two instances of the learning program, denoted as *Alpha* and *Beta* competed against each other. Both players used linear evaluation function consisting of some number of elementary features (there were 16 of them on each side) chosen from a larger, predefined set of available features (the whole pool of features consisted of 38 elements). The most important term - the piece advantage - was calculated separately and was not altered during training.

After each move Alpha was modifying the set of atomic features actually used by the evaluation function (by replacing the unimportant features with new ones picked from the list of remaining features) and was updating their weights based on the following heuristic: For each encountered board position Alpha was saving its estimated score, which was compared with the backed-up score calculated based on the lookahead procedure. The difference between these values (mainly the sign of this difference) guided the process of exchanging the terms and updating the coefficients in the evaluation function of Alpha. Beta was using a fixed form of the evaluation function through the entire game without altering the components or their weights.

After each game, if Alpha won (or was judged to be ahead of Beta), its set of weights was given to Beta and the training continued. Otherwise, Alpha would receive a warning and whenever Alpha received an arbitrary number of such warnings (this limit was usually set to three) it was assumed that the development of Alpha is on the wrong track and the weight system was significantly disturbed by reducing the largest absolute weight value in the evaluation function to zero. This way the program was moved out of the local minimum of the multi-dimensional search space. Occasionally Alpha was pitted against human players in order to test its playing capabilities.

Learning by generalization was quite successful when applied in the middle-game phases - where it was estimated as playing at a better-than-average level - but failed completely in learning how to play in the opening and endgame phases.

Samuel continued to develop the program [279] by implementing several refinements as well as major improvements compared to the 1959's version, achieving finally approximately a just-below-master level of play.

Another well-known example of successful TD-learning application in self-play regime was an impressive achievement of Gerald Tesuaro's backgammon program described in chapter 6.1. TD-Gammon applied a combination of

TD(λ) training method with nonlinear, neural network-based approximation of the evaluation function. Learning was essentially relying on backpropagating the TD errors obtained from the games played by the learning program against itself.

More recent successful experiments include, above all, the TDL-Chinook, which proved a comparably high efficacy of TD-learning in both self-play and external teacher-based training, leading in both cases to a world-championship caliber program (see section 6.3.2 for more details).

An interesting example of self-play training is the NeuroChess system introduced by Sebastian Thrun [323], which combines self-play with the so-called *explanation-based neural network learning* (EBNN) [228, 324].

The role of EBNN is to speed up the training process by allowing better generalization. This goal is accomplished by defining a separate neural network, called the *Chess model*, which represents the expert domain knowledge, obtained through training based on 120 000 grandmaster games, prior to the experiment. The network is a one-hidden-layer MLP with 165 hidden neurons, and its 175 inputs and 175 outputs are numerical representations of board features (defined by a human expert) of the current position and the position expected after the next two half-moves, respectively.

The evaluation function in NeuroChess is represented by a neural network with 175 inputs (bearing the same meaning as inputs/outputs of the EBNN network), between 0 and 80 hidden neurons, and 1 output representing position evaluation (between -1.0 and 1.0).

The challenge of Thrun's approach is to learn the evaluation function (i.e. weights of a neural network) with TD algorithm, based solely on the final outcomes of the training games. Training is a mixture of grandmaster games-based learning and self-play.

Although NeuroChess never reached the level of play of GNU Chess (being its test opponent) defeating it only in about 13% of games, the experiment pointed out some important advantages and weaknesses of TD-learning based on the final games' outcomes. First of all, NeuroChess's ability of playing openings was very poor, which was the consequence of gradually increasing inaccuracy of position estimation from the final position backwards to the opening one. Another characteristic feature of NeuroChess's play was mixing very strong moves with schoolboy mistakes, which according to Thrun happened quite frequently.

The main conclusion from Thrun's work is the relevance of self-play training. According to the results, learning based solely on observation of grandmaster play (TD-learning in here) is not efficient enough and may lead to several artifacts in agent's evaluation function. An example of such inefficiency is the tendency of NeuroChess (when trained without self-playing) to move its queen into the center of the board in the early stage of the game. This behavior was learned from grandmasters' games, but the program was unable to observe that grandmasters make such moves only when the queen is safe from being harassed by the opponent. In other words the basic idea of

EBNN, i.e. using domain knowledge for finding *explanations* for a given set of examples in order to generalize based on them, is not sufficient in the game of chess since some moves cannot be fully explained based exclusively on the domain theory accessible to the system, *ergo* cannot be properly learned.

9.1.2 Training with External Opponents

Encouraged by the success of TD-Gammon, other researchers followed the idea of using TD methods in self-play mode, which - in several cases - ended with undisputable successes. On the other hand, there are also examples of failures with applying TD self-learning, in which the authors promote the use of external teachers instead. A widely-known example that supports relying on external trainers is KnightCap - the TDLeaf(λ)-based chess playing system developed by Baxter, Tridgell and Weaver - described in section 6.3.1. One of the reasons of self-play inefficiency in this case might be a large number of parameters (around 6 000 in the latest version of the system) to be tuned in the evaluation function.

The preference for playing against external opponent rather than learning by self-play was also suggested by Schraudolph et al. [293, 295], where the authors compared the results of TD-learning against *r-Wally* (a version of public domain Go program Wally diluted with random play) or against the *Many Faces of Go (MFoG)* levels $2 - 3$ (stronger than Wally) with the outcomes of playing against itself. The best results were obtained when training was performed against stronger program (MFoG), followed by the ones after training against r-Wally. The self-play mode appeared to be the weakest way of training, although it should be noted that test opponents were the training programs, i.e. r-Wally and MFoG, which gave some preference for the respective learners (the same training and testing program).

One of the first systematic, critical comparisons between self-play and external opponents-based training was Susan Epstein's work on HOYLE - an *advisors*-based multi-game playing system[3].

In the paper under the telling title "Toward an Ideal Trainer" [95] Epstein elaborates on the inherent limitations of the training performed against one, particular opponent. She writes: "The fragment of the problem space that the learning program experiences is influenced by the places to which its trainer leads. After exposure to a limited portion of the search space, the learner is expected to function as if it were an expert in the entire space." Consequently, training with one opponent, whether in a self-play mode or against the expert player is insufficient. In the self-play mode only part of the search space can be properly learned, due to the limited exploration capabilities of the players. A similar conclusion is valid, when training takes place against expert or even against an infallible player. Note that learning how to play efficiently

---

[3] Although the training regime proposed by Epstein is different from TD-learning, her findings regarding the efficiency of various training schemes are general and to a large extent applicable also in TD methods.

against a perfect player does not imply achieving a similar competency level against weaker opponents, since the latter may lead the game into completely unknown positions which the learning program has never had a chance to explore, due to its trainer's perfection in the game. Hence, some diversity needs to be introduced into the trainer's decision system in order to make the training "universally" effective. Diversity of the trainers allows the player to learn how to exploit the flaws in the opponent's play. This ability, according to Epstein, is a central part of the so-called *game player's Turing test*. In order to develop a flexible and "intelligent" player (i.e. the one capable of passing the Turing test in games) it is necessary to confront it with a large range of playing styles and various playing skills.

One of the intuitive solutions to the narrowness of single-opponent training is adding some randomization (noise) into decision process of the perfect or highly experienced trainer. According to Epstein, noise can be implemented in the system in the following way: for $(100 - e)\%$ of time the trainer makes the best move (or chooses one of them at random in case there is more than one such move) and for the remaining $e\%$ of time the move is selected randomly. In the above description $e\%$ can be interpreted as the percentage of the situations in which the trainer has the opportunity to make a mistake (in the case of "randomizing" the infallible trainer) or - more generally - the percentage of moves in which the trainer has the possibility to diverge from the principal variation path. Such an approach, as discussed in [95], broadens the fraction of the search space explored by the learning agent, however, it still does not guarantee that the training is sufficiently universal and that the part of a state space visited by the learner is representative enough to allow efficient play against intelligent, nonconventional opposition.

Consequently, it is proposed to apply *lesson and practice (l&p)* learning instead, which consists in interleaving the periods of training with strong opponents (the lessons) with periods of knowledge consolidation and usage (the practice). The above scheme was successfully applied in HOYLE, but only to relatively simple games like tic-tac-toe, lose tic-tac-toe, or nine men's morris. The question of its applicability to more demanding games has not been addressed.

The assumptions underlying HOYLE's design and learning mechanisms are discussed in more detail in chapter 14, in which the problem of efficient multi-task learning in games is considered. Despite the universality of Epstein's conclusions one ought to be careful with straightforward application of the l&p idea in TD-learning framework. The fundamental difference between HOYLE's and TD's principles is that the former is a type of informative learning where explicit reasons for particular moves are provided in the form of advisors' constructive voting. This type of learning seems to be better fitted to l&p approach than the TD-learning. Nonetheless, on a general note the diversity of the opponents seems to be crucial in achieving a high playing competency, regardless of the training strategy used.

The key role of the choice of the opponents was recently pointed out in TD-GAC experiment, which confirmed the intuition that training with too strong or too weak opponents may not lead to expected improvement since weak opponents simply play badly and the strong ones are too good to be followed by the learner. Also the training scheme, when playing against external opponents, may have a great impact on the speed and quality of the learning process. In particular, in TD-learning one may consider updating weights of the evaluation function after each game or only after the games lost or drawn. Another possibility is to update the weights regardless of the game's outcome, but with elimination of weak moves which most probably may be misleading for the training process [18, 19]. One may also consider playing with stronger opponent a few times in a row if the learner keeps losing against that opponent. The results for the game of give-away checkers presented in [213, 242, 243] suggest the superiority of the above stronger-opponents-focused approach over classical TD-learning based on either all games played or only the ones not won by the learner. Other constructions of the learning scheme, e.g. tournament choice of the opponents, can also be considered. The issue of how to define the optimal training scheme deserves further investigation and hopefully new conclusions across various game domains will come into light.

9.1.3 Human vs. Artificial Trainers

In majority of the TD-learning systems discussed in this book the role of the opponent was played either by the program itself or by other computer programs. High popularity of computer-supported training is caused mainly by its flexibility (the experiment can be designed according to virtually "any" scenario), instant accessibility of the opponents (as opposed to playing against humans) and controllability of the learning progress (the increase or decrease of the learner's skill compared to the trainer is easily measurable). On the downside, training is limited by the playing strength of particular trainer and, what's more important, by the specificity of the machine trainer's playing style.

Considering the above limitations, the possibility of training with humans is a tempting alternative, mainly due to potential diversity of their way of playing. Training against humans is particularly advantageous when they do not realize that their opponent is a computer player and do not try to adapt their playing style. Another relevant factor is a large number of potentially accessible trainers. Both these postulates can be fulfilled when, for instance, training takes place on the Internet gaming zone.

Some insights into discussion on employing human vs. artificial trainers came from the TDL-Chinook experiment. Although both training modes (self-play and playing against Chinook) led to development of world-class checkers programs, examination of weights assigned to particular features in the course of training revealed some relevant discrepancies between trained programs and the hand-tuned Chinook system. The biggest surprise was the difference between weights assigned to the "king trapped in the corner" feature. The TD-learning-based programs found this feature as being of relatively lesser interest than indicated by the human-designed Chinook's set of weights. The reason was much higher frequency of occurrence of this factor in human-vs.-machine play than in the machine-vs.-machine competition. As explained by Schaeffer et al. [288], historically machines were vulnerable to having the king trapped in the corner, and people often attempted to exploit this weakness. Chinook, on the other hand, almost never tried to take advantage of such possibility and consequently, in the course of training, this feature occurred very seldom, usually not being the key factor.

The above observation suggests that optimal training results might be achieved in some kind of interleaved training in which both human and computer trainers are involved. Such approach is strongly advocated by Billings et al. [31] with respect to poker. Although their neural network-based system described in chapter 11, was not trained in TD manner, but with backpropagation learning, the observations concerning the slowness and limited adaptability of self-play computer training presented in the paper are to a large extent universal. As stated in [31]: "Even with a carefully selected, well-balanced field of artificial opponents, it is important to not over-interpret the results of any one experiment. Often all that can be concluded is the relative ranking of the algorithms amongst themselves. One particular strategy may dominate in a self-play experiment, even though another approach is more robust in real games against human opponents. ... For this reason, playing games against real human opponents is still indispensable for proper evaluation."

Generally speaking, the experiences in computer-vs.-computer and computer-vs.-human training collected to date lean towards the conclusion that the most effective training is a blend of both approaches. It seems also reasonable to say that training (or at least extensive testing) against humans is highly desirable if the target playing environment is that of human players.

9.1.4   Exploration-vs.-Exploitation Dilemma

One of practical problems in design and implementation of TD-learning mechanisms is related to finding appropriate balance between *exploration* of unknown areas of the state space (leading to new experiences and to development of new ideas) and *exploitation* or reinforcement of already possessed knowledge.

On the one hand, an agent should keep visiting already known states several times in order to strengthen the tendency to prefer advantageous states and avoid the disadvantageous ones. Only by repeating these experiences the appropriate selection mechanisms can be effectively reinforced. On the other hand, the agent needs to search for new solutions, which may turn out to be more effective than those learned so far. Exploration of new states is also indispensable for proper generalization of agent's knowledge, which in turn leads to increasing agent's flexibility and its ability to play against a wider spectrum of opponents.

The easiest way to implement the exploitation mechanism is application of the *greedy policy*, which in each state assumes choosing the best possible move (according to the current agent's evaluation function). If $s_1, s_2, \ldots, s_k$ denote all possible moves from the state being considered, s^{next} is the move selected by the policy, and $P(s)$ denotes the heuristic evaluation of state reached by move s, then

$$s^{next} := argmax\{P(s_i), i = 1, \ldots, k\} \tag{9.1}$$

In case there is more than one choice, a selection among the candidate moves that maximize $P(\cdot)$ in equation (9.1) is made at random with uniform probability.

The main drawback of strictly following the greedy policy during training (especially in self-play mode) is the imbalance between exploration and exploitation. Playing always the best possible move results in repeatedly traversing the same paths in the game tree, with little chance to explore new, previously unseen game positions. Hence, in practice the greedy policy is usually avoided in self-play training mode and is executed only in the competitive test games.

A strictly greedy approach can be modified towards better exploration capabilities by allowing an agent to choose a weaker move according to some probability distribution. Although such randomization increases exploration abilities, it, at the same time, hinders possible comparisons between trained agents with respect to their playing strength (as they don't always choose the best possible move). The above nondeterminism leads also to low repeatability of results. It is generally agreed, however, that the above inconveniences are negligible compared to possible gains resulting from implementation of a more efficient training process.

The two most popular methods of move selection are ε-*greedy* and *soft-max* policies. In the ε-greedy policy the agent executes the best possible move in all cases except for ε fraction of them. More precisely the value of $\varepsilon \in (0,1)$ denotes the probability that a move in a given situation will be chosen at random among all possible moves, other than the best one. Formally, if

$$s_m = argmax\{P(s_i), i = 1, \ldots, k\} \tag{9.2}$$

then

$$s^{next} = \begin{cases} s_m & \text{with probability } 1 - \varepsilon \\ s_i(i \neq m) & \text{with probability } \frac{\varepsilon}{k-1} \end{cases} \tag{9.3}$$

In a slightly modified implementation, the random choice does not exclude the highest-scored move, i.e.

$$s^{next} = \begin{cases} s_m & \text{with probability } 1 - \varepsilon\frac{k-1}{k} \\ s_i(i \neq m) & \text{with probability } \frac{\varepsilon}{k} \end{cases} \tag{9.4}$$

In some approaches ε represents the frequency (not the probability) of the above described random choice.

In the *softmax policy* [314], instead of choosing randomly with the same probability for selecting each move other than the best one, the agent chooses randomly among all available actions, but according to some probability weighting system. Actions which are expected to yield higher rewards have higher probabilities of being chosen than actions for which the agent expects lower rewards.

The most popular implementation of softmax policy follows the Gibbs (Gibbs-Boltzman) distribution [127], where the probability of selecting move $s_i, i = 1, \ldots, k$ is proportional to $e^{P(s_i)}$:

$$s^{next} := s_i \quad \text{with probability} \quad \frac{e^{\frac{P(s_i)}{T}}}{\sum_{j=1}^{k} e^{\frac{P(s_j)}{T}}}, \quad i = 1, \ldots, k \quad (9.5)$$

In equation (9.5) parameter T is called *temperature*, and its role is to model the relative importance of the best move compared to the other ones. For "high" temperatures all probabilities are close to each other and approximately equal to $\frac{1}{k}$. For "low" temperatures, the probability distribution is more focused on the best move and the remaining choices of moves have relatively lower probabilities. In the limit of $T \longrightarrow 0$ in (9.5) the method approaches the greedy policy.

Usually in the initial phase of training high values of T are used, which allows wide exploration of the state space (though with relatively low playing quality). As the training progresses the temperature is slowly decreased and the exploration gradually gives way to exploitation. Finally, at temperature close to zero strategy (9.5) approaches the greedy one. The above described scheme of decreasing the temperature in time is known as the *simulated annealing* procedure [128, 168] and used as a powerful global optimization technique. In the domain of mind games the method was mainly applied to Go playing programs [34, 294, 295].

Another example of softmax policy implementation was presented in [220], where the so-called *creativity factor* c has been proposed. In this method a move is selected randomly according to uniform distribution, among all the moves whose evaluation fits the interval $[P_c, P_{max}]$, where P_{max} equals the assessment of the best move and P_c is defined in the following way:

$$P_c = P_{max}(1 - c), \quad 0 \le c \le 1 \quad (9.6)$$

In practical applications the impact of the softmax-creativity method changes in time. In the initial stage of training, due to evaluation inefficiencies, relatively more moves can fit the desired interval and have a chance of being selected (with equal probability). In a more advanced training phase, the best move can more often be the only one that fulfils the criterium of being estimated not lower than P_c.

Yet another possible approach to alleviation of the exploration-vs.-exploitation tradeoff is application of a multi-opponent-based training instead of a single-opponent-based one. In a multi-player environment, exploration can be enforced by appropriate selection of trainers with no need to abandon the greedy policy. The major factors contributing to the successful balance between exploration and exploitation, when multi-trainer TD-learning schemes are used, concern the following issues [213, 242, 243]:

a) the number of trainers,
b) trainers' playing strength,
c) conditions under which the current trainer is replaced by another one,
d) conditions under which the game played by the agent is taken into account in the training process.

The above issues are thoroughly discussed below in chapter 9.2 devoted to various TD-learning strategies.

9.2 TD-Learning Strategies

Different choices referring to points a)-d) listed above lead to different training schemes within the TD-learning paradigm. Several examples of such training strategies are presented below in this chapter. All of them were implemented and verified in the TD-GAC experiment.

9.2.1 Learning on All Games

The most straightforward implementation of a training scheme in $TD(\lambda)$ or $TDLeaf(\lambda)$ learning is unconditional modification of weights after each game played against the trainer, regardless of the game's outcome. This strategy (denoted here by LB) is usually applied in the initial phase of training especially in the case when the weights of the agent's evaluation function are initialized randomly. Frequent weight updates allow making relevant changes based on relatively small number of games, i.e. within a short training time.

Once the training process advances these frequent weight changes become less favorable and in some situations may cause the opposite effect of deterioration of the learning agent's playing abilities, due to reasons discussed below with description of strategy LL.

9.2.2 Learning Exclusively on Lost or Drawn Games

The next strategy (denoted by LL) is generally more effective than LB. Here, the weights are updated based exclusively on games lost or drawn by the agent. Losing a game undoubtedly shows that the agent was weaker (in this particular game) than his opponent. Therefore the lesson learned by this game ought to be memorized by the agent, which should avoid playing the same game positions in the future.

Winning a game by the agent does not provide such a clear indication. The success may in this case be either caused by high playing competency of the agent or by poor play of his opponent. In the former case the tendency to play according to the same policy should be strengthened, but in the latter case adaptation of weights towards the weak opponent should be avoided - a stronger opponent (encountered in the future) would most probably exploit these flaws. The problem of attributing the reasons for winning the game to either strong play of the agent or poor play of the opponent is one of the basic dilemmas in TD-learning methods. In strategy LL looking for a solution to this problem is avoided by arbitrarily assuming that in case of winning the game by the agent updating of weights is too risky and should not be done. In the experiments carried out by Mańdziuk and Osman [213] this strategy appeared to be superior to LB in case of partly trained agents.

9.2.3   Learning on "No Blunders"

An important disadvantage of strategy LL in comparison with LB is smaller number of games which are used to change agent's weights. Games won by the learning player are useless in this process. In order to alleviate this problem a modification of strategy LB, denoted LBNB (NB stands for "no blunders") was proposed. In this strategy all moves of the opponent which are not classified as blunders are used in the learning process - regardless of the final result of the game.

Opponent's move at time $t + 1$ is assumed to be "no blunder" if it was either causing a decrease in the learning agent's position evaluation (was a strong move) or, otherwise, was predicted by the agent as a move to be played, during the tree search process performed at time t. In other words, the opponent's move was lying on the path with the highest minimax score (*principal variation path*). This idea is depicted in figure 9.1, where the opponent's move, which increases the evaluation of the learning player's position, is classified as "no blunder" if $s_{t+2} = s'_{t+2}$.

A similar idea was proposed earlier by Baxter et al. in KnightCap, where there was no weight update associated with moves not predicted by the learning player (treated as blunders) in case the learning player outplayed a lower ranked opponent.

9.2.4   Learning with Focus on Strong Opponents

Another strategy, denoted by L3 [242, 243], modifies LL in the following way: after the game is lost by an agent, his weights are updated in the same way as in LL, but the next game is played against the same training opponent, if only the number of games lost in a row against that opponent is less than three. Otherwise (after three consecutive losses, or a victory, or a draw of an agent) the opponent is replaced by the next trainer.

The underlying idea of this method is giving the learning agent a chance to play more games against the opponents stronger than itself. At the same time

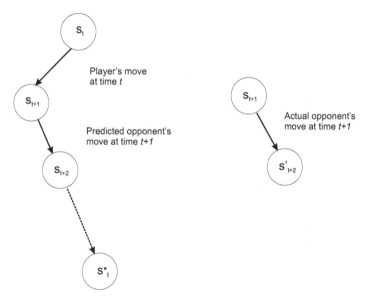

Fig. 9.1. Prediction of the opponent's moves in strategy LBNB. The move, which increases the position value of the learning player is not considered to be a blunder, if it is lying on a principal variation path (leading to node s_t^* in the figure) found in the search process. This is equivalent to fulfilling the condition $s_{t+2} = s'_{t+2}$

there is no risk that the player would follow any particular trainer too closely thanks to the restriction of playing at most three games in a row with the same trainer. This limit can be adapted to particular circumstances depending on the nature of the game, availability of the training opponents and the advancement of the learning process. In TD-GAC experiment the value of 3 appeared to be well chosen, though no rigorous optimization procedure was performed with regard to this selection. In domain of give-away checkers, strategy L3 clearly outperformed strategies LL, LB and LBNB [243] and was comparable to strategy LR (presented below).

9.2.5 Successive Elimination of Weaker Opponents

All strategies mentioned so far utilize a fixed population of training opponents. The order of trainers is usually defined according to some preestablished permutation or trainers are chosen at random with uniform probability.

Strategy LR (Learning with Removal) proposed in [241] differs in the way in which a population of the opponents is maintained during training. Here the population is not fixed and at each epoch the worst performing trainer is left out. The trainers are chosen for playing according to some predefined permutation and each trainer maintains his overall score against the agent. After every n completed training epochs, for some n, the trainer which collected the lowest cumulative score against the agent is removed from the set of trainers and omitted in subsequent games. The training continues with the same order of the opponents (without the one left out). The process is terminated when there is only one trainer left. As an effect of gradual elimination of weaker trainers, the average playing strength of the opponents the agent is confronted with, increases in time.

The results obtained with applying strategy LR are very competitive among tested strategies (see chapter 9.3). The critical factors in the LR method are the choices of n and the initial number of trainers. In general, the more opponents involved in the training process the more universal the evaluation function developed by the agent. Furthermore, the bigger n the more stable the learning process. Both these parameters, however, have a significant impact on the time required for the learning procedure to be completed.

It should be noted that due to decreasing diversity of the opponents the ability of exploration of the state space decreases in time, and the balance between exploration and exploitation gradually changes in favor of the latter one.

9.2.6   Random Opponent

One of the interesting possibilities for increasing the exploration capabilities of the learning agent is training against random opponent. Such a player uses a random evaluation function while searching the game tree. Whenever the game tree is searched to the leaf position, a true value of the actual outcome, i.e. MIN, MAX or 0, is returned according to (7.1). If the state is nonterminal, $V(s, w)$ in (7.1) is a $random()$ function, which returns a random integer from the interval $[MIN + 1, MAX - 1]$. Hence, definition (7.1) is replaced by the following one:

$$
P_{rand}(s, w) = \begin{cases} MAX \text{ if } s \text{ is a win} \\ MIN \text{ if } s \text{ is a loss} \\ 0 \text{ if } s \text{ is a tie} \\ random() \text{ for all other } s \in S \end{cases} \tag{9.7}
$$

Random opponent searches the game tree only in order to look for the terminal states. The advantage of applying such a randomly playing teacher is the possibility of entering game positions, which would have been hard to reach otherwise, i.e. in "normal" play.

Observe that in the literature a random player is sometimes defined as one with randomly chosen weights in the evaluation function or the one which chooses a move at random with no search whatsoever. Certainly, each of the above players exhibits significantly different behavior than a random player defined in this section. In the remainder of the book, in order to avoid any possible misunderstandings a *random* player will denote one which behaves according to equation (9.7), whereas the *pseudo-random* player will refer to the one with randomly chosen weights in the evaluation function[4]. A player that chooses a move randomly, without any search, will not be considered.

9.3 TD-GAC Experiment

In section 6.3.3 an overview of the TD-GAC experiment was presented emphasizing the simplicity of the technical means applied: weighted linear evaluation function composed of 22 features only, common for all game stages and shallow tree search with depth $d = 4$. In this chapter the experiment is described in detail focusing on its two main goals: comparison of the selected training schemes implemented within $TD(\lambda)$ and $TDLeaf(\lambda)$ frameworks, and practical verification of the training strategy consisting in playing against several different sets of gradually stronger opponents. The training was divided into three stages: training against pseudo-random opponents, then against strong players, and finally against very strong ones.

In the first training phase, when the learning entries were initialized with random weights in the evaluation function, the $TDLeaf(\lambda)$ method appeared to be significantly inferior to $TD(\lambda)$. Based on these preliminary results it was decided to use exclusively the $TD(\lambda)$ in the initial training and include the $TDLeaf(\lambda)$ only in subsequent stages. Also a more sophisticated strategy - L3 was not applied in the first stage, since L3 is dedicated to fine-tuning of the learning agent's playing skills rather than to learning from scratch. The learning methods tested in TD-GAC experiment were compared with each other and additionally with a simple $(1+1)ES$ - a hill-climbing technique denoted by EVO.

The EVO Method

EVO relies on local search in the neighborhood of the opponent's weight state. The method was proposed by Pollack and Blair [258] and, with some minor modifications, by other authors afterwards [174, 243]. In this method the learning agent and his opponent use the same form of the evaluation function, being usually a linear combination (7.1) of game features, differing only by the associated sets of weights.

The way the method was implemented in TD-GAC experiment is schematically depicted in figure 9.2, where $PLAY(pl_a, pl_p)$ denotes playing one game between players pl_a and pl_p (the agent and the trainer), in which the player (pl_a) makes the first move. Function $SHIFT(pl_a, pl_p, m)$ moves the weights of player pl_a towards the ones of player pl_p by the amount of $(pl_p - pl_a)m$. Function $ADD\text{-}WHITE\text{-}NOISE(pl_p)$ changes the weights of player pl_p by adding random noise $N(0, 1)$ independently to each weight.

---

[4] Certainly, from a formal point of view, none of the two above-defined players is *truly* random, as far as computer implementations are concerned.

function EVO()
1:   pl_a = learning agent;
2:   pl_p = opponent;
3:   **repeat**
4:        result = 0;
5:        result += PLAY(pl_a, pl_p);
6:        result += PLAY(pl_p, pl_a);
7:        **if** ($result < 0$)
8:             SHIFT($pl_a, pl_p, 0.05$);
9:        pl_p = ADD-WHITE-NOISE(pl_p);
10: **until**(stop_condition);

Fig. 9.2. EVO - a hill-climbing, coevolutionary algorithm. The pseudocode comes from [241] and refers to the case of perfect-information, deterministic games

In the experiment the learning agent plays some number of games against his opponent with sides swapped after each game. The results are scored $+1, 0, -1$ for each agent's win, draw, and loss, respectively. If the agent achieves a negative score in a two-game match against the trainer his weights are changed towards the ones of the trainer by the amount of 5%, since in such a case it is assumed that the trainer's weight vector is more efficient than that of the agent. After every two games the trainer's weight vector $w_1, \ldots, w_n$ is modified by adding a random value u_i from the $N(0, 1)$ distribution to the element $w_i, i = 1, \ldots, n$, independently for each i. Certainly, due to random modifications of the trainer's weights, the direction of agent weights' improvement also changes in time.

The implementation presented in fig. 9.2 is tailored to deterministic games with greedy strategy, and therefore only two games are played between the player and the opponent with sides swapped after the first game. In the case of nondeterministic games, or deterministic ones played with non-greedy policy, the number of games played between the learning player and its opponent may be greater than two. Also the benchmark defining the clear advantage of the opponent over the player can vary in time. Such implementation was proposed originally in [258] for backgammon. The details are presented in chapter 9.5.

Stage 1. Training with Pseudo-random Opponents

Recall that in the TD-GAC experiment there were 10 learning agents, 8 of which were initialized with random weights, and the remaining 2 had their weights initialized with all zeros.

In the first stage of the TD-GAC experiment the learning players were trained by 25 pseudo-random opponents. This stage has already indicated the efficacy of the TD-learning methods. Both TD-LL and TD-LB visibly exceeded the 50% threshold, which means that the learning policy was quite successful in this phase. The highest average result (72.2%) was obtained using the LB method during games 5 001 − 7 500. In subsequent games the performance decreased. Results achieved by the EVO method were 6 − 8% worse than those of TD strategies.

A summary of results is presented in Table 9.1. The entries in the table reflect the average performance of all 10 learning agents obtained by saving their weights after every 250 training epochs (totally there were 400 such players; 10 000/250 = 40; 40 · 10 = 400). Test results against unknown pseudo-random and strong opponents are presented in Table 9.2. In the case of pseudo-random opponents no qualitative differences can be observed between TD-LL and TD-LB strategies, however both of them are visibly more effective than the EVO method. Moreover, the performance of TD-LL and TD-LB players against pseudo-random opponents practically does not decrease between the training and testing phases, which supports the claim about generality of this type of learning. In the case of EVO method the performance degrades by a few per cent points. Tests against strong opponents, show clear 1 − 8 per cent point superiority of TD-LL over TD-LB, and further advantage over the EVO, whose performance rapidly deteriorates against strong opponents.

Observe that the results of training are not monotonically increasing and several significant fluctuations can be observed. The learning curves of the best TD-LL and TD-LB players are presented in figures 9.3(a) and 9.3(b), respectively. These results were sampled after every full round of 25 training games. The best overall score (24.5 out of 25 points) was achieved by player 9-LB between the 6 000th and 8 000th game (fig. 9.3(b)). The worst overall result (not presented) was a drop to 13 points by player 7-LL.

Both curves in figure 9.3 are not smooth and sudden raises and drops can be frequently observed, especially in the initial training phase in the case of TD-LL strategy. These sudden performance changes are the main obstacle in defining appropriate stopping conditions, which are hard to be developed either *a priori* or online, based on the analysis of the learning process. A reasonable approach is to define a threshold level, exceeding which causes the training process to terminate. Alternatively, one may continue training as long as time permits and choose the best player among all developed in the entire process.

Table 9.1. Results of training with pseudo-random opponents in stage 1

games	TD		EVO
	LL	LB	
1 − 2 500	64.9%	64.7%	64.5%
2 501 − 5 000	70.5%	70.2%	63.5%
5 001 − 7 500	72.0%	72.2%	64.8%
7 501 − 10 000	70.8%	69.1%	63.1%

Table 9.2. Results of the test phase after stage 1 training

games	Pseudo-random opponents			Strong opponents		
	TD		EVO	TD		EVO
	LL	LB		LL	LB	
$1-2\,500$	66.8%	69.9%	51.7%	44.2%	43.9%	30.1%
$2\,501-5\,000$	70.1%	70.9%	55.7%	51.6%	43.2%	32.9%
$5\,001-7\,500$	71.7%	70.0%	57.2%	49.2%	43.5%	35.3%
$7\,501-10\,000$	70.7%	70.7%	59.4%	50.0%	44.5%	36.8%

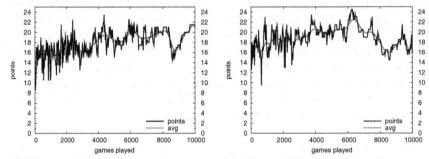

(a) Player 8-LL. Pseudo-random oppo-   (b) Player 9-LB. Pseudo-random opponents.   nents.

Fig. 9.3. The learning curves of the best TD-LL and TD-LB players tested against unknown pseudo-random opponents

Stage 2. Training with Strong Opponents

After playing 10 000 games against pseudo-random opponents, the learning players were trained by 25 strong opponents during a period of another 10 000 games. The results of training are presented in Table 9.3. In this stage the TD-LL method was visibly superior to TD-LB. The difference in performance varied between 9 and 20 per cent points in all training phases. Application of the TD-LB training resulted in slightly exceeding the level of 50%. The best overall performance (71.9%) was observed with the TD-LL method in the third training phase (between games 5 001 and 7 500).

Table 9.3. Stage 2 training phase. L3f denotes strategy L3 with a fixed value of α. EVO player was continuing training according to the algorithm presented in fig. 9.2

games	TDLeaf		TD				EVO
	LL	LBNB	LL	LB	L3	L3f	
$1-2\,500$	61.3%	59.0%	61.8%	52.9%	56.7%	56.7%	65.3%
$2\,501-5\,000$	66.3%	65.6%	64.3%	54.9%	45.5%	45.5%	65.7%
$5\,001-7\,500$	60.1%	63.3%	71.9%	51.8%	40.8%	44.5%	62.2%
$7\,501-10\,000$	60.1%	64.6%	62.8%	52.3%	37.5%	44.0%	70.1%

The worst training results were accomplished by TD-L3 method. Over time the learning players started to lose more and more games against the trainers. The reason was gradual decreasing of step-size α in subsequent games[5]. At this stage, when playing against strong opponents, it appeared that applying one or two weight modifications (after the first and the second consecutively lost games) was usually not enough to win the third game. This observation was especially true when small learning step size was used. Hence, in most cases the learning player recorded 3 losses in a row against the same opponent. Fixing the value of α resulted in 4 - 7 per cent points improvement compared to the case with α decreasing in time. On the other hand, the TD-L3 training results are generally expected to be inferior to other methods, since TD-L3 agents deliberately play more games against opponents stronger than themselves.

In strategy LL, the TDLeaf results were similar to TD ones, except for exceptionally high TD-LL's score during games 5 001 − 7 500. The TDLeaf-LB strategy was enhanced by preventing the learning player from considering the opponents' blunders. This restriction was applied only when the learning player had won with the opponent. In the case of a loss or a tie, all encountered states were taken into account (cf. description in section 9.2.3). Due to this enhancement the TDleaf-LBNB method was not inferior to TDLeaf-LL, as was LB to LL in the case of TD. TDLeaf-LBNB method showed stable performance, without sudden changes observed in TDLeaf-LL learning. Moreover, TDLeaf-LBNB was clearly superior to TD-LB.

The EVO results presented in Table 9.3 were obtained as a direct continuation of the previous experiment (stage 1). Due to different learning strategy applied in EVO there was no opponent change at the beginning of stage 2. Overall, the results accomplished in stage 2 were slightly better than those of stage 1.

Results of the stage 2 test phase are presented in Tables 9.4 and 9.5. The best performance against strong opponents was accomplished by TD-L3, which slightly outperformed TD-LL and TD-L3f. The next place belonged to TDLeaf-LBNB. TD-LL method appeared to be visibly more effective than TD-LB in the case of strong testers. EVO algorithm once again turned out to be the weakest choice.

Table 9.4. Stage 2. Tests with pseudo-random opponents

	TDLeaf		TD				EVO
games	LL	LBNB	LL	LB	L3	L3f	
1 − 2 500	71.8%	71.9%	72.0%	69.5%	73.5%	73.5%	58.7%
2 501 − 5 000	70.6%	72.5%	74.4%	71.2%	74.9%	74.2%	56.2%
5 001 − 7 500	70.4%	72.4%	72.5%	72.1%	72.8%	73.2%	56.8%
7 501 − 10 000	70.2%	71.4%	73.1%	73.2%	74.4%	72.1%	54.9%

---

[5] Gradual decrease of α over time, as discussed in section 9.4.5, is indispensable for successful implementation of other TD-learning methods.

Table 9.5. Stage 2. Tests with strong opponents

games	TDLeaf		TD				EVO
	LL	LBNB	LL	LB	L3	L3f	
1 – 2 500	51.7%	50.4%	53.3%	43.3%	53.3%	53.3%	35.2%
2 501 – 5 000	52.2%	55.0%	56.7%	44.8%	57.4%	56.1%	33.4%
5 001 – 7 500	50.6%	56.1%	54.0%	47.9%	56.7%	56.4%	32.8%
7 501 – 10 000	50.8%	56.0%	54.8%	48.2%	56.1%	55.8%	31.6%

Stage 3. Training on Yet Stronger Opponents

In the third stage training was continued for another 10 000 games. The goal was to check whether the results would improve any further. In general, the answer was negative. Most probably the highest possible performance that could be achieved with a simple value function and shallow search has been reached. Numerical results referring to this stage are presented in [243].

Tests of TD-LR, TD-LBNB and TD-L3NB Methods

Some preliminary tests were also performed with strategy TD-LR described in section 9.2.5. The results were encouraging, and even better than those of TD-L3. In three out of five experiments test results against strong opponents (the ones used in stage 2 testing) exceeded 57%, with the highest score being equal to 59.5%, which is the best overall result[6]. The best result was accomplished with $\alpha = 1E-5, \lambda = 0.5$, the initial number of opponents equal to 28, and the number of training games equal to 10 125. In the tests with pseudo-random opponents the learning player scored 74%.

The TD-LR strategy seems to be worth further consideration and possibly after some refinements may lead to statistically stronger training method than TD-L3. One of the possible modifications is replacing the currently weakest trainer with a new one picked from some predefined set of trainers, thus not decreasing the number of opponents in time and avoiding gradual hindering of the game space exploration. In the case of training with replacement the cumulative ranking should be substituted by the single-epoch-based one.

The superiority of TDLeaf-LBNB over both TD-LB and TDLeaf-LL suggests that applying a combination of TD and LBNB strategies, may cause the improvement over the basic TD-LB method. This intuition was confirmed by numerical tests, in which the combined strategy TD-LBNB achieved the score of 58.4% against strong opponents, thus outperforming TD-LB (48.2%), TD-LL (56.7%) and TD-L3 (57.4%), and only slightly losing to the best TD-LR result (59.5%).

Another potentially promising idea is to join the strength of LBNB learning with TD-L3 method, yielding the TD-L3NB strategy. The efficacy of this

---

[6] It should be noted, however, that according to Wilcoxon test, the difference between TD-LR and TD-L3 was not statistically significant.

combination was not confirmed in practice. Although adding the LBNB modification to L3 allowed considering all played games in the training process (not only the lost ones as in the original formulation of L3) the combined strategy TD-L3NB performed worse than TD-L3. Apparently, a straightforward combination of two successful heuristics does not always lead to qualitatively better solution (but certainly is worth trying!).

9.4 Lessons from TD-GAC Experiment

The TD-GAC experiment provided quite an extensive amount of data supporting conclusions about strong and weak points of particular TD-type training methods. The four main, general level aspects worth emphasis are the following:

- the choice of the training strategy - in particular the choice of games used for training;
- the selection of training opponents - with preference to multi-stage design of the training process where the strength of trainers progresses at each step along with the improvement of the learning program;
- the role of random opponent in the training process - pros and cons of applying this type of training;
- problems with convergence.

The above issues have been already discussed in previous sections of chapter 9 as an immediate illustration of the effects of applying particular training scenarios. In the following sections these findings are recapitulated in a systematic way.

9.4.1 The Choice of Training Strategy

One of the main conclusions drawn from the TD-GAC experiment was confirmation of correlation between the choice of the training strategy and the quality of results. Based on the Wilcoxon test [345] statistical relevance of the differences between tested methods was verified. According to that test the following two hierarchies of methods, from the weakest to the strongest one, can be established: when tested against strong opponents - EVO; TD-LB; TDLeaf-LL; (TDLeaf-LB, TD-LL); (TD-L3, TD-LR, TD-LBNB)[7]; when tested against pseudo-random opponents - EVO; TDLeaf-LL; (TD-LB, TDLeaf-LB); (TD-LL, TD-L3, TD-LBNB, TD-LR). The main difference between these two statistics is exchanging positions between TD-LB and TDLeaf-LL and between TD-LBNB and TD-LR.

In general, the most effective are strategies in which learning takes place based exclusively on the games lost (TD-LR and TD-L3), or lost and drawn (TD-LL) by the agent. Learning based on games won by the agent may

---

[7] Strategies in each of the parenthesis are not statistically differentiable.

be misleading if the success is caused not by proper play of the learning program, but by poor play of its opponent. In the latter case the learning system often becomes "too optimistic" and its assessment of position is inadequately high compared to its objective value. In other words the agent tends to overestimate positions which may possibly lead to victories only after weak opponent's play. This tendency was clearly observed when analyzing particular games played by the trained agents, and also reflected by a few per cent difference in accomplished results between TD-LB and TD-LL in favor of the latter one when tested against strong opponents, capable of exploiting weaknesses in agent's play.

On the other hand, a clear advantage of LB strategy when compared to LL is the possibility to use all experimental data, which in turn causes more frequent weight update. This feature is useful in the initial phase of the training, when the learning program starts with randomly assigned weights. An interesting and effective solution is combining "the best of both worlds", i.e. applying TD-LBNB strategy. Similarly, the TDLeaf-LBNB learning approach appeared to be both strong and stable.

9.4.2   Selection of Training Opponents

Another key issue when applying TD methods in a multi-player environment is the number of opponents involved in the training process. In typical TD experiment the opponents are chosen in a cycle according to some predefined permutation or selected at random with uniform probability. Generally speaking the greater the number of trainers the better. Diversity of the opponents assures that playing abilities of the agent are general and not focused on particular opponents. However, the greater the number of trainers the longer the training process and the more difficult the convergence. Moreover, as observed in the TD-GAC experiment, increasing the set of opponents above some (unknown beforehand) threshold value does not improve the results anymore. Tests with 100 trainers ended-up with weaker performance of the learning agent than in the case of 25 trainers, after the same number of training games.

During the experiments it was also observed that training against many opponents allowed the agent to break the initial tendency to draw majority of the games, which was the case when training was arranged with one locally modified opponent. Diversity of the opponents was also indispensable for attaining qualitatively higher level of play in the test rounds with previously unknown test players. These improvements were attributed to better generalization which was the effect of better exploration of the state space, which in turn was reflected in the evaluation function developed by the learning agent.

This is why it is strongly suggested in [213, 242, 243] to use a relatively large group of training players with diverse evaluation functions, representing various playing styles. Such approach provides an appropriate balance between exploration and exploitation and was proven experimentally to be more effective than playing against one locally modified opponent (strong domination of exploitation) or against a random opponent (strong domination of exploration).

It is also recommended to divide the training process into several stages and use gradually stronger, on average, sets of opponents in subsequent training phases. This method allows steady improvement of the playing skills of the learning agent, yet, this strategy has its inherent limits as well, and at some point further improvement is not possible without more profound changes or developments in the training methodology.

9.4.3   Training and Testing with Random Opponent

In theory, the advantage of using random opponent described in section 9.2.6 should be twofold. First of all, playing with such an opponent may lead the game into specific positions very rarely encountered otherwise, thus promoting exploration of the game space and enhancing generalization abilities of the learning agent. Second of all, such an opponent may be used to verify the playing strength of the learning player by making comparisons at various search depths. Since the greater the search depth d the higher the chances for a random player to reach the end-of-game state and return the true estimation of a given state, such a comparison allows assessment of agent's evaluation function against the brute-force search.

In TD-GAC experiment, the random opponent did not appear to be an effective trainer. Relatively poor training results were most probably caused by high imbalance between exploration and exploitation. Too many new, weakly related states reached during training games did not let the player focus on regularities of the game and learn useful, generally applicable playing policy. Observe, that random trainer may return a completely different estimation of the same game position each time it is assessed. Consequently, training with random opponent may be regarded as training with divers, practically non-repeatable players. On the other hand, the simplicity of the training process (shallow depth of search, plain evaluation function, one set of weights used in the entire game) must have also been contributing to the final outcome of training. Perhaps, with more sophisticated training environment, longer training time and carefully-tuned schedule of decaying the learning parameters in time, the process might converge to a more successful result.

Application of a random opponent as a test player was influenced by the results of Jonathan Schaeffer and his students' experiment, which aimed at verifying the efficacy of give-away checkers randomly playing agent that used deep search depths - up to 15 ply (plus extensions). One of the conclusions was that deeply-searching random player is hard to beat even by specially trained programs equipped with sophisticated evaluation functions [285].

These observations were confirmed in TD-GAC experiment. When both TD-GAC and random player used shallow search of $d = 4$ the TD-L3 method was able to score 68.0%, but as the search depth of random player was increased (with TD-GAC's fixed at $d = 4$) the results deteriorated to 35.8% and 18.1%, respectively for depths of 6 and 8 [241]. The reason for such high scores of the random opponent lies in its inherent tendency to play nonstandard paths and explore new game territories. Both the agent and the random opponent possessed only very fragmentary, incomplete knowledge about the true values of these unknown states. In such cases deeper search was an important asset.

Due to extreme variability of decisions, the random player is actually nonpredictable, which makes a significant difference compared to the player equipped with randomly chosen (though fixed) set of weights. Consequently, random player appeared to be a clearly stronger opponent for TD-L3 method than a collection of 100 pseudo-random testers (recall that the TD-LL (TD-L3) test result against these pseudo-random players was equal to 71.7% (74.9%) after the first (second) training phase, respectively). The superiority of random player was also confirmed in direct test games. In four independent experiments, each against a set of 100 pseudo-random players, a random player achieved the following results: 57.3%, 59.8%, 58.1%, and 60.1% .

One of interesting directions that may lead to stronger play against random opponent (assuming that the player knows the nature of its opponent) is playing in a more risky way based on the assumption that since the opponent plays randomly it will not necessarily exploit potential weaknesses of a non-perfect, but promising move. Consider for example the case in which the player in state s has two possible moves leading to positions s_1 and s_2, respectively, and in position s_1 9 out of 10 possible continuations lead to the victory of the player and the remaining one to a loss, whereas position s_2 always leads to a draw (see figure 9.4). A player following the minimax playing policy would chose continuation s_2 not taking the risk of potentially losing the game after playing move s_1. However, the expected return from playing s_1 is higher than from playing s_2, since

$$E(s_1) = 0.9, \qquad\qquad E(s_2) = 0.5 \qquad\qquad (9.8)$$

assuming standard scoring system $(1, 0.5, 0)$ for win, draw, loss, respectively. Generally speaking, if the agent followed a policy of maximization of the expected return then its score against random player would most probably increase[8].

---

[8] Certainly, this reasoning is valid only if search depth is insufficient for reaching the leaf nodes of a full game tree. Otherwise (in end-of-game positions) the true leaf-state values would be taken into account by a random player.

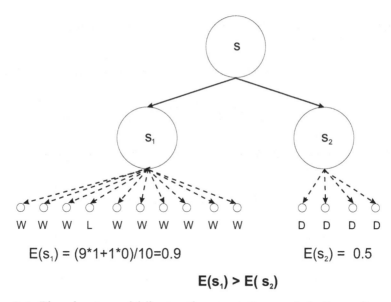

$$E(s_1) = (9*1+1*0)/10=0.9 \qquad E(s_2) = 0.5$$

$$E(s_1) > E(s_2)$$

Fig. 9.4. The advantage of following the expectation maximization policy when playing against random opponent. Nodes denoted by W, L, D represent the leaves of the paths leading to win, loss, draw, respectively. Conservative player, which adopts the greedy minimax policy would play s_2 as an optimal choice - always leading to a draw. More aggressive player might play s_1 - as indicated by the expected return

9.4.4   Oscillations in the Learning Process

The agents' performance during TD-GAC training was not monotonically increasing and, as can be seen in the exemplar figures 9.3(a) and 9.3(b), their progress graphs were full of local oscillations. There are two main reasons for such instability of training results.

The first one is attributed to the nature of TD-learning. After each game the agent changes its weights in order to, first of all, become more effective against the opponent it has been just playing with. These modifications may cause degradation of agent's playing abilities against other players. This phenomenon, known as *a moving target* problem, often occurs in supervised training methods, e.g. in backpropagation neural network training [100]. Naturally, with appropriate initial choice of learning coefficients, and suitable schedule of their decay in time, one may expect the process to stabilize and find the resultant direction of weight changes allowing overall improvement of the learning system. This assumption is valid, but only if the complexity of the TD-learned function is adequate to that of the real, approximated function. A linear combination of 22 plain features is apparently too simplistic to guarantee the convergence of the training process in the case of give-away-checkers - a game with presumably nonlinear value function.

The other reason is connected with the goal of the training process. The TD-learning method used in TD-GAC attempts to approximate a real value function in the entire game state space, whereas in practice it is sufficient to only be able to only compare the states (one is better/worse than the other). Such a goal is significantly simpler than an attempt to learn approximate values of all possible game states. Following an example presented in [241], assume that in a certain game position s the agent has two choices, denoted by s_1 and s_2, respectively, and that the real values of these states are respectively equal to $+2$ and -1 (see figure 9.5). Consider two cases. In the first one (denoted by c_1) the agent, having the weight vector w_1, assigns the values $V(s_1, w_1) = 30$ and $V(s_2, w_1) = -10$, and in the other one (c_2) the respective values are equal to $V(s_1, w_2) = -1, V(s_2, w_2) = 1$, for some weight vector w_2. The mean square errors are equal to

$$MSE_{c_1} = \frac{1}{2}\sqrt{865}, \qquad MSE_{c_2} = \frac{1}{2}\sqrt{13}, \qquad (9.9)$$

but even though $MSE_{c_1} \gg MSE_{c_2}$, it is the choice c_1 (i.e. move s_1), that is correct in this case. In the above example it is sufficient (and more beneficial) to learn that state s_1 is more advantageous than state s_2 (i.e. $V(s_1, \cdot) > V(s_2, \cdot)$), without attempting to approximate the exact values of these states.

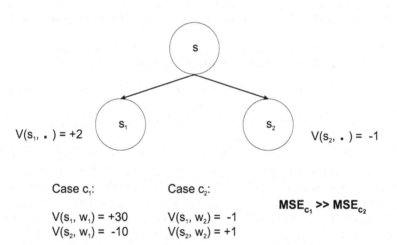

Fig. 9.5. Learning numerical approximation of the goal function and learning the relative assessments of games states (which of the two states is higher valued) are not always equivalent goals. In the figure the evaluation function with the weight vector w_1 (case c_1) yields higher mean square error in the task of numerical goal function approximation than the function with weight vector w_2 (case c_2), but nevertheless leads to appropriate choice of the next state s_1, as opposed to the other case, in which state s_2 is selected

9.4.5   Tuning α and λ

One of the conclusions from TD-GAC experiment is that parameters α and λ in (5.15) and (5.16) need to be decreased gradually along with the learning progress. Appropriate choice of initial values and schedules of changing these parameters in time are crucial for achieving good performance in both TD(λ) and TDLeaf(λ) algorithms. Gradual decrease of of the step-size parameter α was essential for obtaining satisfactory training results whereas suitable tuning of the decay parameter λ substantially shortened the learning time, e.g. the maximum performance could be reached after 7 500 training games compared to over 10 000, in the case of fixed λ. One of the possible schedules of decreasing α and λ proposed in [213] is presented in Table 9.6. This scenario was applied in all TD-GAC experiments discussed in this chapter.

Table 9.6. The schemes of decreasing α and λ in time

games	α	λ	games	α	λ
1 − 2 500	1E-4	0.95	2 501 − 5 000	2E-5	0.70
5 001 − 7 500	1E-5	0.50	7 501 − 10 000	5E-6	0.20

9.5   TD-Learning vs. Evolution

Both evolutionary methods and TD-learning represent very popular approaches to developing game playing agents. In the last 30 years substantial game research was performed within these two frameworks. Both domains have been quite deeply reviewed based on theoretical analysis and experimental results, but despite the evident increase of knowledge concerning these methodologies there are still several basic questions that are not convincingly answered. One of them refers to potential gain of using one of these techniques over the other.

The above question can be partly answered by analyzing the results of applying both methods to learning the same tasks (games). Some indicative examples of such comparisons are reviewed in the remainder of this chapter.

The first, "classical" example is backgammon, where Tesauro's work [317, 318] can be compared with Pollack et al.'s coevolutionary approach [258, 259]. The authors of the latter work applied the EVO method to the task of developing a competitive backgammon player implemented as a 4 000-weight feed-forward neural net, using the same game representation as did Tesauro. The evolutionary process started with the initial champion player having all weights set to zeros, playing against its slightly modified copy. The modification consisted in adding some Gaussian noise to the original player. The champion and the challenger played a certain number of games and whenever the number of challenger's wins exceeded a predefined threshold the new champion was defined by applying the following *parent-child averaging* scheme with $\beta = 0.05$:

$$w_i' := (1 - \beta)w_i' + \beta w_i, \quad i = 1, \ldots, 4\,000 \tag{9.10}$$

where w_i', w_i denote the ith weight of the champion and the challenger, respectively. The rule (9.10) appeared to be far more effective than simple replacement of the champion by the successful challenger. The authors found it also beneficial to apply some kind of annealing procedure to steer the number of wins required for the challenger to influence champion's weights. In the beginning of the coevolutionary process the challenger was required to win at least 3 out of 4 games, after 10 000 generations the bar was raised to 5 out of 6 games, and after 70 000 generations it was raised again to the level of 7 out of 8 games. Applying this simple scheme for 100 000 generations led to development of a relatively strong backgammon player capable of winning about $40 - 45\%$ of test games against *Pubeval* - a publicly available, moderately strong backgammon program.

Pollack and Blair estimated the playing strength of the evolved player to be comparable to Tesauro's 1992 TD-player. In the response paper [320] Tesauro disagreed with the statement that coevolutionary system accomplished a comparable level, recalling that TD-Gammon's result against Pubeval was equal to 56% with the same neural architecture as the one of [258]. Moreover, Tesauro argued that Pollack and Blair's coevolutionary approach failed to learn the nonlinear structure of the problem and despite the presence of nonlinear hidden neurons was only capable of linearly approximating the optimal backgammon playing policy. In effect the policy learned in coevolutionary regime was not competitive to "truly nonlinear" evaluation function acquired by TD-Gammon.

The above discussion was continued and formalized by Darwen [76], who compared application results of coevolutionary and TD methods in backgammon, in the universal context of the No Free Lunch (NFL) theorems [349]. The NFL theorems state that the method applied must suit the problem to be solved. The question of the *practical applicability* of linear vs. nonlinear game representation in the coevolutionary and TD approaches is an immediate consequence of these theorems.

Darwen found that for linear evaluation functions a coevolutionary technique is capable to outperform the TD-learning method, although the time required to evolve a high-performing individual is substantially longer than in the case of TD-learning. In the case of nonlinear evaluation functions the time requirements increase dramatically, which makes the coevolution less effective than TD methods. Darwen estimated that for nonlinear architectures, if the number of games required for successful TD-learning equals few hundred thousands, then in order to attain a similar learning result with coevolutionary method billions of games are necessary. The reason for this discrepancy is attributed to different ways of approaching the final solution in both learning frameworks. The TD incremental update is computationally a much lighter method than population-based coevolutionary approach, in which, at each generation, a comparison between any two individuals requires a sufficient number of games to be played. Darwen concluded that in the case of backgammon and nonlinear form of the evaluation function the computational load of coevolutionary process was prohibitive.

Yet another comparative studies between coevolutionary and TD methods were performed by Kotnik and Kalita, who reported in [174] an advantage of EVO approach over the TD method in a card game *gin rummy*. Gin rummy is a moderately complex game, and its search space consists of approximately $2 \cdot 10^{31}$ states. On the other hand, the branching factor of gin rummy is not high, and the game is not strategically demanding - it is played based on a simple set of rules and with a small set of possible actions at each turn [174].

Kotnik and Kalita closely followed Pollack and Blair's approach. In particular they applied the same type of noise when defining the new challenger and used a parent-child averaging scheme (9.10). Experimental results were in favor of the coevolutionary method, which was able to come up with more comprehensive playing strategies than the TD method. The authors suggested that the reason for weaker performance of the TD approach were inefficient exploration mechanisms. This diagnosis is likely to be true considering that TD-learning was performed in a self-play regime with greedy move selection policy. Due to the above marked inefficiency of TD implementation, it is hard to judge the results. In order to make the results conclusive, an application of ε-greedy (or other non-greedy) policy in TD training would be advisable. Apparently, the amount of randomness coming from the random cards' distribution was insufficient to ensure effective exploration.

Kotnik and Kalita's results are inconsistent with the findings of TD-GAC experiment where clear advantage of TD and TDLeaf methods over EVO in both preliminary and advanced phases of training has been observed. One of the possible explanations of this discrepancy lies in a very careful design of the TD training procedure in TD-GAC, as opposed to [174], where TD method was applied in a very basic way.

An interesting and insightful comparison between coevolutionary learning and gradient-descent TD(0) was presented by Runarsson and Lucas [275] for small-board 5×5 version of Go. In both methods the evaluation function was represented in the form of the WPC (7.5) with *board_size* $= 25$ and $i = 0, \ldots, 25$. Value of s_0 was set to 1 and represented a bias term, $s_i, i = 1, \ldots, 25$ represented board's intersections, with values equal to $+1$ for a black stone, -1 for a white one, and 0 in the case of empty intersection. Furthermore, $V(s, w)$ in (7.5) was squashed to fulfill $\mid V(s, w) \mid \leq 1$, by applying hyperbolic tangent function, i.e. $V(s, w) := \tanh(V(s, w))$. Since the values of the intersections were determined by the game position, the learning task consisted in defining the set of 26 weights $w_i, i = 0, \ldots, 25$.

In the TD method the weights were updated during the game using a standard gradient-descent implementation [314]:

$$w_i := w_i + \alpha[r + P(s', w) - P(s, w)]\frac{\partial P(s, w)}{\partial w_i} \qquad (9.11)$$

where s and s' denoted respectively the state in which a player was to move and the state after executing the chosen move. If s' was a terminal state, then $P(s', w) = 0$. The value of r equaled 0 during the game, and only in the terminal state became a *reward* (+1 for the win of black, −1 for the win of white, and 0 for a draw). α was a learning rate coefficient, which had to be appropriately decreasing in time. During learning the ε-greedy policy was applied, with $\varepsilon = 0.1$. Different initialization schemes were tested with respect to weights $w_i, i = 0, \ldots, 25$. The best results were obtained when all weights were initialized with zeros.

The coevolutionary learning (CEL) process was tested with various choices of steering parameters: population size λ, elitist $(1 + \lambda)$ versus non-elitist $(1, \lambda)$ replacement schemes, noise-including vs. noise-free policies, and several weight initializations in (7.5). Moreover, following Pollack and Blair [258] the *parent-child averaging* (9.10) was applied with $\beta = 0.05$. The strength of mutation was either fixed or adapted according to the so-called *geometrical averaging* scheme [275].

The choice of β, the question of using fixed vs. self-adaptive mutation scheme, and the selection of initial weights w were yet another (besides algorithm's steering parameters) possible choices of the CEL implementation.

Various settings of the above control parameters of the CEL were tested in [275]. Based on these choices and the settings of the TD algorithm a number of experimental comparisons between these two learning paradigms were made leading to interesting insights into each of these methods separately, and also showing the relative strengths and weaknesses of one with respect to the other.

The general conclusion was that coevolutionary process, when carefully tuned (suitable choice of the replacement scheme, population size, number of generations, introduction of noise to the policy) outperformed the results of TD(0) gradient-descent training, although the latter converged to a solution much faster - approximately an order of magnitude fewer training games were needed for TD. In the design of coevolutionary process the parent-child averaging scheme (9.10) appeared to be particularly important for both elitist and non-elitist replacement strategies. Moreover, the elitist strategy was slower to learn than the non-elitist one. The results confirmed also that a higher level of play can be achieved in CEL by increasing the population size (from 10 to 30 in the experiment), at the cost of nearly tenfold increase in the number of games played at each generation (from 45 000 to 435 000)[9]. Additional improvement in CEL was accomplished by adding noise to the move selection policy during evolution.

---

[9] In each generation each individual competed with each other individual by playing two games (with sides swapped) receiving the score of $+1, 0, -2$ for win, draw, loss, respectively.

The main observation concerning the TD method, except for its faster convergence, was the requirement for appropriate decay of α in time in (9.11). The authors concluded that suitable selection of this parameter's schedule was critical and proposed using the following scheme: initialize α with 0.001 and reduce it by $\frac{2}{3}$ every 4 500 games. This scheme was superior to using fixed value of $\alpha = 0.001$. The observation concerning the relevance of suitable choice of α is in accordance with similar conclusions drawn by Mańdziuk and Osman in TD-GAC, and also by other researchers running TD experiments.

In 2006 Lucas and Runarsson conducted analogous experiment in the domain of Othello [199] and reached very similar conclusions. Again the TD method allowed much faster learning, but was slightly inferior to properly tuned coevolutionary process. The authors underlined that standard coevolution performed very poorly and that parent-child averaging again appeared to be essential for getting satisfactory performance.

One of the key issues reported in both Go and Othello related papers was that the complexity of the function to be learned should be tailored to the problem being represented. The simplicity of WPC representation and only one-ply minimax search used in the experiments hindered its ability to learn/represent more sophisticated game concepts, e.g. the concept of an *eye* in Go[10].

The simplicity of the WPC-based approximation had also an impact on the TD-learning convergence. This form of evaluation function appeared to be too simplistic to properly model the (possibly highly nonlinear) real dependencies. In effect, the TD process tended to oscillate and the attempt to stop the algorithm after a certain, fixed number of iterations was unsuccessful. Instead the authors proposed monitoring the learning process and then choosing the best performing set of weights [199]. Recall that the same observations were true in the TD-GAC experiment where the authors also failed to find automatic procedure for detecting the end-of-simulation conditions.

Despite the above criticism related to the limitations of the simple position representation, it should be underlined that the coevolutionary approach proposed by Lucas and Runarsson is an example of knowledge-free learning system, in which human intervention is restricted exclusively to setting-up the CEL internal parameters and defining the conditions of competitive promotion to the next generation.

Prior to the above described research, Lubberts and Miikkulainen [194] aimed at experimental comparison between the coevolution and the evolutionary training with a fixed Go player, also on a 5×5 board, but with the evaluation function represented by a neural network. The evolutionary process was organized according to the Symbiotic Adaptive Neuro-Evolution (SANE) method [232, 267]. The authors concluded that coevolution was

---

[10] Naturally, a small size of the board also introduced some limitations to representation of complex, strategic concepts, and to some extent led to oversimplification of the learning process, but for representation of the basic ideas of both games, a 5×5 board seems to be large enough.

faster and more effective than evolution with a fixed opponent (one of the earlier versions of GNU Go in that case) and in particular avoided the problem of hindering the population development by the limited playing abilities possessed by a single, invariable opponent.

The substantial number of research results concerning TD and coevolutionary learning methods presented in the literature provide significant insights into potential capabilities and inherent limitations of both learning frameworks. There are, however, still some topics which are not convincingly addressed, and in the author's opinion are worth consideration in the quest for new research developments.

Apart from various detailed questions concerning each of these training techniques alone, partly discussed above in this chapter, one of the especially tempting fields, which has not been extensively explored yet, is hybridization of both methods. Having known the general pros and cons of temporal difference and coevolution it is in theory possible to design new training methods that combine the best properties of TD (fast convergence, low computational load) with those of coevolution (fine refinement of the final solution), while trying to avoid their weak points. It will be interesting to observe the prospective development of this promising, almost uncharted research area.

One of the best ways in which methodology-related open issues can be researched and verified is direct comparison of players developed with various techniques and learning strategies under competitive tournament conditions. A great opportunity for such verification is Computer Olympiad organized each year by Jaap van den Herik and his collaborators (the recent ones have been located in Amsterdam (2007) and Beijing (2008)).

Besides the Computer Olympiad there are also other contests organized in conjunction with major CI-related conferences. In particular, in the last few years, the idea of driving the game research forward by means of competitions and tournaments was intensively pursued by Simon M. Lucas and Thomas P. Runarsson, who have been organizing and running the Othello Competition [198], which offers a chance to test various neural architectures in Othello domain.

Standard MLP architectures, spatial Blondie24-type networks, as well as WPC networks have been widely tested in previous editions of this competition. A very interesting recent extension is the possibility to test the efficacy of n-tuple networks. N-tuple networks were proposed initially by Bledsoe and Browning [35] in the late 1950s with regard to optical character recognition, and recently rediscovered by Lucas for the game playing domain [195, 196]. The architecture of an n-tuple system allows simultaneous sampling of the input space with a set of n points. Assuming that each sampled point can take one of m possible values, the entire sampled entry can be treated as an n-digit number in base m, and used as an index into the array of weights [195]. Such a construction allows flexibility in defining weights for particular sampled entries (depending on the entries' values), but at the same time requires relatively high number of weights (despite partial unification of weights that can be imposed based on symmetries and rotations of the board). In the case of Othello, n-tuple networks trained with TD method proved to be a better choice than standard MLPs or WPC architectures, and in particular outplayed the CEC 2006 Champion by Kim et al. [167] - the best network submitted to date for the Othello Competition [195]. Further development of n-tuple systems and especially their application to other mind games seems to be an interesting research durection.

Regarding the postulate of possible exploration of hybrid TD-learning and coevolutionary approaches, it is worth to note, that the above-mentioned CEC 2006 Champion resulted from a synergetic cooperation of coevolution and TD-learning. The authors coevolved the MLP architecture seeding the population with some variations of a well performing MLP obtained previously with the TD-learning.

The success of their approach suggests that when the evaluation function is represented by a neural network it may be worth investigation, also in games other than Othello, to use TD-learning for initial, crude estimation of the evaluation function and then apply evolutionary techniques to the population defined in the neighborhood of this preliminary solution.

Move Ranking and Search-Free Playing

Quick and effective ranking of moves based on their rough, initial estimation can be used either in a stand-alone mode, i.e. in a search-free playing program, or as a training tool for human players, or - most frequently - as a preliminary step before more expensive analysis is performed. In particular, initial sorting of possible moves according to their potential strength is often used in alpha-beta pruning algorithms, thus leaving more time for exploration of the most promising paths. Certainly in competitive tournament play, due to time constraints, efficient move pre-ordering must be either implemented in a search-free way or rely on shallow search in order to devote the remaining time to deeper, selective search.

Efficacious move ranking is a very intuitive, "human" skill. Human chess players can, for example, estimate roughly two positions per second - compared to 200 billion ones verified by Deep Blue in the same time - hence being extremely effective in preliminary selection of moves.

Methods of initial move ranking can be roughly divided into three main categories: methods considering the strength of a given move in various positions in previously played games (mentioned in chapter 10.1), methods relying on the existence of particular game patterns in a given position (presented in chapter 10.2), and methods considering the existence of predefined game features in a given position (discussed in chapter 10.3). All three types of methods take advantage of some expert knowledge about the game, in the form of either relevant patterns/features or the heuristic assessment of a general usefulness of a given move. Pattern-based methods are particularly useful in Go, which has an obvious characteristics of a territory game and in which configurations of stones play the leading role in assessment of a board position.

Naturally, these three major categories do not exhaust the list of potential methods. In particular, there is a number of specific approaches dedicated to chess, that have been developed through years in the spirit of the Shannon's type-B algorithms. Two recent ideas of move pre-ordering in chess domain are presented in chapter 10.4. The first one relies on metrical properties of a

J. Mańdziuk: Knowledge-Free and Learning-Based Methods, SCI 276, pp. 155–168.

chess position, the other one considers a heuristic assessment of the potential move's impact on particular sections of the board.

The ordering of moves can also be included in the process of the evaluation function learning. One of the possibilities is to use the cost functions that impose higher penalty for moves that are low ranked than for the top-ranked ones in case they are selected as one of the best moves [218]. This procedure is expected to ultimately lead to the cost function that is carefully tuned towards selection of the highest ranked moves and ignores the negligible ones.

10.1 Methods Relying on Historical Goodness of a Move

There exist a few popular, generally applicable search-free strategies of move ranking relying on historical goodness of a move in previously played games. These strategies include the *history heuristic*, the *killer move heuristic*, *transposition tables*, or *refutation tables*, discussed in section 3.2.3. In majority of game situations these methods are very effective, since the assumption that a move which often appeared to be efficient in the past, is more likely to be a strong one also in the current game position than other "less popular" moves, is generally correct. The above heuristics are widely applied to enhance the alpha-beta game tree search, whose effectiveness highly depends on the appropriate ordering of moves in each node. This issue has been thoroughly discussed in chapter 3.2 devoted to game tree searching algorithms.

10.2 Pattern-Based Methods

Pattern-based approaches are perfectly suited for territory games, such as Go. However, due to limited pattern analysis abilities exhibited by computers, current Go programs rely only on simple pattern matching. Inability to process board patterns effectively is one of the main impediments in building world-class Go playing systems, which would eventually become a serious threat to top human players. In this context, application of pattern-oriented methods of move pre-selection seems to be a promising research direction in this game. This avenue was underlined by Martin Müller in his excellent overview paper [237]. One of the most challenging problems in Go domain was therein formulated as: "Evaluate and empirically test different approaches to pattern matching and pattern learning in Go. Generalize to more powerful (human-like?) patterns." Design of pattern-oriented move ranking mechanisms is one of possible realizations of this claim.

One of the interesting recent approaches, based on using pattern-templates of predefined shapes, was proposed by Stern et al. [308] in Go. A pattern was defined as a particular configuration of stones within a template centered at the empty location, on which a move is to be made. The set of patterns was generated automatically by analyzing game records and choosing the frequently encountered ones. Following the work of van der Werf et al. [330] discussed in the next section, the predictive system was enhanced by adding to each pattern eight predefined binary features describing crucial tactical information including the existence of *Ko*, the number of *liberties*, or the distance to the edge. Based on 181 000 training games the system learned probability distribution of the moves' values in a given position according to the context provided by the local pattern. Such a distribution, as stated in the introduction to this chapter, can be used as a stand-alone Go playing program or as a move pre-selection tool prior to the regular search.

The system was tested on 4 400 games played by professional players. In 34% of test positions the system pointed out the move which was actually played by the expert during the game, in 66% of them the expert move was within the top-5 system's recommendations, and in 86% of positions in the top-20 moves ranked by the system.

Stern et al.'s idea of using pattern-templates was subsequently employed by Araki et al. [8] in another Go move ranking program. The authors additionally proposed to use the 64-bit Zobrist hashing method [354] allowing for significant time and memory savings, partly due to reducing the representation of all symmetric instances of a given pattern to a single hash value. The move ranking model was implemented in two steps. First, for each training position all legal moves were ranked according to the relative frequencies of the template-patterns matching the stone configuration. Next, the top n ranked moves were used in the re-ranking process based on Bayesian estimation of the goodness of a given move with respect to the existence of particular features.

Several experiments aimed at tuning system's parameters were performed and reported in [8]. In the final test on 500 master games from the **Games of Go on Disc** (GoGoD) database [145] the best move was ranked the first in 33.9% of test positions, and in 60.5% and 69.5% of them the best move was within the top-5 and the top-10 choices, respectively. The above figures are slightly inferior to Stern et al.'s ones, but were obtained based on as few as 20 000 training positions only.

10.3 Methods Using Predefined Game Features

Another popular group of move ranking methods consists of feature-based approaches. Usually such algorithms are focused on verifying whether given features are present on the board and on calculating the numerical values of these features whenever applicable. In the game of Go examples of such features include the number of player's stones on the board or a distance to the board's edge. In the simplest case the move ranking system can be built in the form of a linear weighted combination of the features' values. In the case of more elaborate solutions some type of learning (supervised or unsupervised) is involved. Generally speaking, the feature-based approaches to move ranking problem consist in constructing appropriate evaluation function of a game position which can be efficiently applied without search.

An interesting example of learning-based method that considers predefined features in the game of Go is presented in [330]. The authors are focused on local move prediction with the help of neural network training. The MLP architecture with one hidden layer composed of one hundred units with hyperbolic tangent transfer function is used as local move predictor. The underlying assumption is that the next move will be executed locally - within a certain neighborhood of the previous move, called region of interest (ROI) in the paper. Such an assumption is generally true for Go, due to strong locality of the game. Several kinds of ROIs differing by shape and/or size were tested, leading to the conclusion that the size matters much more in this case than does the shape. Hence, in subsequent experiments a fixed shape of ROI, namely diamond, was used.

A set of local, easily computable features was defined as the input for the neural network (local) move predictor. These features included the location of stones within the ROI (in the form of a binary vector), local edge detection, the existence of the ko-rule and, if applicable, the distance between the move to be played and the place of an illegal move due to the ko-rule, the number of liberties for each point within the ROI, the number of liberties that a newly added stone will have after the move is made, the number of stones to be captured in effect of making this move, the Manhattan distance from the last opponent's move to the considered move, and some number of specific features related to the "nearest stones" outside the ROI (see [330] for their detailed description).

In order to improve the quality of a neural predictor a training was performed in pairs, i.e. in a given position two moves were considered - the move actually chosen by an expert player and a random move within a Manhattan distance of 3 from that executed move[1]. Having a pair (M_e, M_r) of training moves, where M_e, M_r denote respectively the expert move and the random one chosen according to the above described procedure, the following error function (10.1) was used:

$$E(v_e, v_r) = \begin{cases} (v_r + \varepsilon - v_e)^2, & \text{if } v_r + \varepsilon > v_e \\ 0, & \text{otherwise} \end{cases} \qquad (10.1)$$

where v_e, v_r denote the predicted values of the expert move and a random move, respectively, and ε is the tolerance coefficient responsible for controlling the desired minimal difference between both moves. Typically, in the applied RPROP training [268, 269] parameter ε belonged to $[0.1, 1.0]$.

In order to improve the results further, the network was subsequently retrained on the set of preprocessed features extracted from the above-described raw input data by the PCA (Principal Component Analysis [161]) transformation.

---

[1] A similar idea was previously proposed by Tesauro in the *comparison training* algorithm [315].

During tests performed using master games played on 19×19 board the system ranked 48% of the actually played moves as its first choice, when restricted to the local neighborhood, and 25% of them in case the full board was considered. In the latter case, in 45% of test cases the move actually made by a professional player was among the top-3 ranked moves, and in 80% of test positions among the top-20 moves. On 9×9 board the move played during the game was selected in 55% of test cases, when selection was restricted to the local neighborhood, and in 37% of them, when the full board was considered. In the full board selection case in 99% of positions the expert move was placed in the top-25 ranked moves.

A brief comparison with human performance on the local move prediction task placed the proposed method at the level comparable to that of "strong [human] kyu-level players" [330].

A combination of feature-based and pattern-based representations was proposed in [312], where *convolutional neural networks* were used as search-free predictors of the best move in Go.

Convolutional neural networks [181] are feed-forward networks with shared weights (c.f. a description in chapter 8.2). Each hidden layer is composed of a few rectangular planes of the same size as the (rectangular) input pattern, which receive signals from the previous layer. In fig. 10.1 a one-hidden-layer convolutional network is is presented, in which the hidden layer is composed of 3 such planes. Each element of these 3 hidden layer's planes receives stimulus from a certain sub-rectangle of the input layer only. The size of the sub-rectangle determines the degree of locality of processed information. The weights from each sub-rectangle are shared and replicated among all hidden layer's planes assuring that each local patch of the input image is proceeded in the same manner.

The basic set of shared weights is called *the convolutional kernel*. Each plane can be considered a feature map with a fixed feature detector (convolutional kernel) since the weights of a kernel are fixed for all points/neurons in a plane. In [312], a square kernel of size $n \times n$ is implemented. The input value presented to a hidden neuron (x, y) is calculated in the following way:

$$inp_{(x,y)} = \sum_{u=-(n-1)/2}^{(n-1)/2} \sum_{v=-(n-1)/2}^{(n-1)/2} s_{x+u,y+v} K_{u,v} \qquad (10.2)$$

where $s_{i,j}$ denotes the value of input element (i, j) and $K_{u,v}$ is weight (u, v) of the convolutional kernel K. In case the convolutional kernel extends beyond the board's edge the elements pointing outside the board are ignored. The output value of a hidden neuron is a sigmoid transformation of its input, i.e.:

$$out_{(x,y)} = (1 + e^{-inp_{(x,y)}})^{-1} \qquad (10.3)$$

In the experiment, the convolutional kernels between the input and the hidden layer were square-shaped of size 7×7 or 9×9. Hidden-to-output convolutional

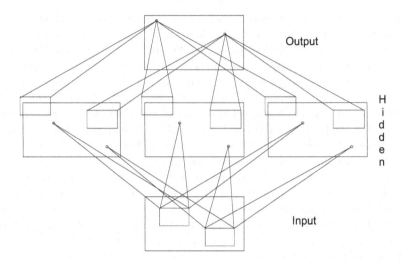

Fig. 10.1. A convolutional feed-forward network with one hidden layer composed of 3 convolutional planes. The weights between input and each of the planes in the hidden layer, as well as between these planes and the output layer are shared

kernels were of size 5 × 5. The hidden layer consisted of 15 convolutional planes.

A board position was represented either in a raw form or in a feature-based manner. In the later case, for each nonempty intersection the number of *liberties* of the intersection's group was calculated and quantified as either equal to 1, or equal to 2, or greater or equal to 3. Combining these three possibilities with two possible colors of a stone placed at the intersection led to six possible outcomes. Hence, the board was represented in the form of six bitmaps of size 19 × 19, each of them devoted to particular combination of the number of liberties and the color of a stone. For each intersection exactly one of these bitmaps had the bit associated to that intersection turned on, the rest of them had this bit switched off. If the intersection was empty the respective bit in all six bitmaps was equal to zero. The output layer consisted of 361 softmax neurons representing the system's belief that a move should be made on the respective intersection.

The system was trained with backpropagation method with momentum. The learning rate was initially set to 0.1 and reduced to 0.01 after 3 000 weight updates. A momentum coefficient was equal to 0.9 during the entire training. The learning proceeded for 100 000 weight updates, where each weight update corresponded to one training game from the GoGoD database. Since the authors used 45 000 training games, the number of completed training epochs was less than three.

The system was further enhanced by adding several local predictors, including the 9×9 template predictors similar to those described in [308] and [8],

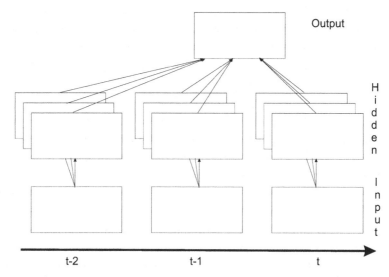

Fig. 10.2. A convolutional network with one hidden layer and convolutional kernels applied to board patterns at different time steps. For the sake of clarity of the figure only three time steps (t, $t-1$ and $t-2$) are presented. In the experiment either one (the current board) or five (t, $t-1$, $t-2$, $t-3$, $t-4$) time steps were used

which yielded a $2-4\%$ increase compared to a raw board representation. The results presented below concern the enhanced input representation case.

In the main experiment two approaches were verified. In the first one only the current board was presented during training. In the other case previously made moves were also considered by presenting boards from four previous time steps together with the current board position - see fig. 10.2. The results indicate that adding the information about the regions in which immediately previous moves were made significantly increases the probability of correct prediction. The best result, equal to 34.6% of perfect predictions, was obtained in the case of applying 9×9 kernel. In the case of 7×7 kernel the performance was slightly worse and equaled 34.1%. Both results were obtained with accompanying presentation of boards from previous time steps. Results obtained based exclusively on the current board presentation were about $12-14\%$ worse. The best overall result (36.9%) was attained in the case of using an ensemble of the above described predictors with appropriate weighting system [312].

10.4   Chess-Related Approaches

The importance of the move ranking task brought about the development of many interesting ideas of how to tackle this problem. Some of them are universal, applicable to various types of mind games, others are dedicated to concrete games. In particular the 60-year history of machine chess playing offers several approaches aiming at partial realization of the Shannon's idea of constructing the type-B programs [296] capable of discarding the unpromising moves automatically, without applying any search or extensive calculations. The most promising ideas are related to pattern analysis or pattern classification tasks since due to relative complexity of chess rules and several nuances in position analysis the rule-based methods are poorly applicable in this domain.

Two recent examples of move pre-ordering methods in chess are presented below in this chapter. Some other, more cognitively plausible systems are discussed in the last part of the book.

Para-Distance-Based Training

A method relying on metrical properties of a game position in chess has been recently proposed by Dendek and Mańdziuk [82]. A set of chess moves E was defined as containing all pairs $x = (x_1, x_2)$ of chessboard positions before and after performing a selected move. Following [158] a metrical space structure was imposed on E by introducing a *para-distance function* $P_{[\alpha,\beta]}$ defined in the following way:

$$P_{[\alpha,\beta]}(x,y) = \sqrt{\alpha(B_1(x) - B_1(y))^2 + \beta(B_2(x) - B_2(y))^2} \qquad (10.4)$$

where α, β are real parameters satisfying the condition $\alpha, \beta > 0$ and functions $B_1(x)$ and $B_2(x)$ are respectively an initial position strength and a position strength change after a move was made, defined in the following way:

$$(B_1(x), B_2(x)) = (b(x_1), b(x_2) - b(x_1)), \qquad (10.5)$$

for some heuristic evaluation function b.

Function P expresses the intuitive notion of a distance between chess moves: two moves are similar (close to each other) if their initial positions' strengths and the moves' effects (the after-move positions' strengths) are pairwise similar.

Note, that P is not a distance function in a strict sense, since

$$\sim (P(x,y) = 0 \Leftrightarrow x = y), \qquad (10.6)$$

which is a consequence of defining P with the use of non-bijective function b.

Definition of P implies that it is possible to divide E into subsets, similarly to division into equivalence classes, $E_i, i = 1, \ldots, k$ such that:

$$\forall_{i=1,\ldots,k} \quad ((|E_i| = 1) \text{ or } (\forall_{x,y \in E_i} \ P(x,y) = 0)) \qquad (10.7)$$

The above property was used in [82] to order the set of training patterns in a neural network supervised training process. Game positions defining those patterns were selected according to the following rules:

- *A move was played in a game between opponents having at least* 2300 *ELO ranking points* (roughly equivalent to FIDE master title, which is just below the international master). This restriction increased a probability that both sides played strong, close to optimal moves.
- *A move was made by the winner of the game.* This postulate again increased a probability that a move performed in a given position was close to optimal, since it was played by the winning side.
- *A move occurred between* 15*th and* 20*th game move.* This restriction was a consequence of proposed classifiers' localization [82], which was not only to the fragment of a chessboard (2D geometrical dimension), but to some sphere including the time dimension. In the future, this set of attributes may be further extended into other functional dimensions in order to decrease the complexity of classification task.
- *A move was made by the side playing white pieces.* This restriction allowed easy distinction between king's and queen's wings in the learning process. If prediction was to be made for black pieces then a position could be reverted by switching colors of the pieces and applying symmetrical reflection of the left and right sides of the board.

Two types of neural network architectures were tested in the experiment: an MLP with one hidden layer containing 30 neurons (denoted by 1H) and an MLP with two hidden layers containing 30 and 20 neurons, respectively (denoted by 2H). In the input layer, every square of the chessboard was represented as a vector over the set $\{-1, 0, 1\}^5$. Each vector's element denoted a particular piece: rook, knight, bishop, king, and pawn. Queen was encoded as a combination of rook and bishop. When a white piece was occupying a given square the respective value was positive; if it was a black one - the respective value was negative. If the square was empty the vector associated with that square consisted of zeros only. A chessboard was represented as a straightforward concatenation of all squares considered in particular, predefined order. Hence the input vector was defined over the set $\{-1, 0, 1\}^{5 \cdot 64}$.

The output part of the training patterns encoded a square of the chessboard on which a move had been carried out in the actual game. Output representation was parameterized by an arbitrarily chosen square $S = g5$. The encoding was defined by a taxicab distance from a target square of a given move to square S, and was represented as an integer value from the interval $[0, 14]$. Consequently, an output pattern was a vector from the space $[0, 1]^{15}$, with zeros on all positions except the one denoting the encoded distance, which value was equal to 1. The equivalence classes defined by the above coding are depicted in figure 10.3[2].

---

[2] Although with the parametrization of the problem relative to square $g5$ the actual output values fitted the interval $[0, 10]$, for the sake of generality the wider range $[0, 14]$ was used, which in the future would allow considering parametrization based on any arbitrary square.

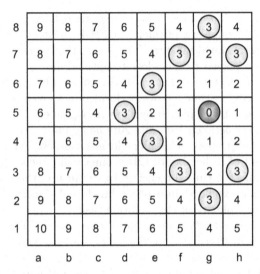

8	9	8	7	6	5	4	③	4
7	8	7	6	5	4	③	2	③
6	7	6	5	4	③	2	1	2
5	6	5	4	③	2	1	⓪	1
4	7	6	5	4	③	2	1	2
3	8	7	6	5	4	③	2	③
2	9	8	7	6	5	4	③	4
1	10	9	8	7	6	5	4	5
	a	b	c	d	e	f	g	h

Fig. 10.3. Equivalence classes in the output space in the training process defined with respect to square $g5$ (denoted by dark circle). Light circles denote all squares belonging to the equivalence class number 3, i.e. with Manhattan distance from $g5$ equal to 3

The set of patterns consisted of 42 401 elements, divided into training set T, $|T| = 38\ 252$ and testing set V, $|V| = 4\ 149$. All positions were sampled from chess games provided in [68]. Training was performed in *interleaved manner* according to one of the training algorithms previously proposed by the authors [81]. The general idea of this method was to interleave the ordered sequence of the training patterns (defined in a specific way described below) with a randomly defined sequence of these patterns. The ordered sequence in the training epoch t was chosen with the following probability:

$$p_t(ord.seq) = p_0 e^{-\eta t} \tag{10.8}$$

where p_0 was the initial probability of ordered sequence selection, and η was a discount factor. The interleaved training yielded significant improvement compared to traditional backpropagation training with one randomly chosen sequence or two random sequences interleaved with each other in the Handwritten Digit Recognition (HDR) domain [81].

In order to define the ordered sequence of the training patterns the following algorithm, called *Model II* in [81], was applied. The para-distance function $P_{[1,1]}(\cdot, \cdot)$ defined by equation (10.4) was used as metric M in the algorithm.

Algorithm of Model II

Given a metric M defined on pattern space and a set of patterns $\{T_m\}$ the average pairwise distance S_n^{II} between the first n elements of the sequence can be expressed as:

$$S_n^{II} = \frac{2}{(n-1)n} \sum_{k=1}^{n} \sum_{l=k+1}^{n} M(T_k, T_l), \tag{10.9}$$

where $M(T_k, T_l)$ is the distance between patterns T_k and T_l. A sequence of q training patterns $(T_l)_{l=1}^{q}$ that fulfils the following set of inequalities:

$$\forall_{1 \leq l \leq q-1} \qquad S_l^{II} \geq S_{l+1}^{II} \tag{10.10}$$

is called *ordered set of model II*.

A given set $\{T_m\}$ composed of q elements can be ordered to sufficiently approximate *ordered set of model II* with the use of the following algorithm:

1. Define any sequence $(T_l)_{l=1}^{q}$.
2. Create an empty sequence O.
3. Create distance array $D[1..q]$:

$$\forall_{1 \leq l \leq q} \qquad D_l := \sum_{k=1}^{q} M(T_l, T_k)$$

4. Choose a maximal value element of D:

$$v := max_{1 \leq l \leq q} \qquad D_l.$$

5. Pick randomly one element r from the set $\{1 \leq l \leq q \,|\, D_l = v\}$.
6. Update distance matrix:

$$\forall_{1 \leq l \leq q} \qquad D_l := D_l - M(T_r, T_l)$$

7. Remove T_r from $\{T_m\}$ and place it at the beginning of the sequence O.
8. Remove D_r from the distance array.
9. Put $q := q - 1$.
10. Repeat steps 4-10 until $q = 0$.

As proposed in [81] the ordered sequence of the training patterns constructed according to the above-defined algorithm was interleaved with a random sequence of these patterns. A standard backpropagation training method with learning rate and momentum equal to 0.7 and 0.1, respectively, was applied. Networks were trained during 1 000 epochs.

For each of the 1H and 2H architectures two types of classifiers were defined: the first one was trained in a standard way according to one randomly selected training sequence and the other one was trained using suggested switching (interleaved) model with the initial probability of switch $p_0 = 1$, monotonically decreasing in time down to 0.03 in the last training epoch.

In order to assess the quality of tested classifiers, a distance between the class predicted by the classifier and the true class has been calculated. Fractions of a population for which the above distance did not exceed certain limits (denoted by *radius*) are presented in Table 10.1. As can be observed

Table 10.1. Percentage of predictions with a distance from correct classification not greater than *radius*. For example, in the case of 2H network trained with sequence switching 35% of the answers pointed the correct class and 64.87% were either correct classifications (the above-mentioned 35%) or pointed the neighboring class (the remaining 29.87%). The maximum error (Manhattan distance from the correct class) in the whole testing set did not exceed 6 squares in this case

	% of population in a sphere			
radius	% *with* sequence switching		% *without* sequence switching	
	1H	**2H**	**1H**	**2H**
0	28.42	35.00	27.31	34.51
1	52.66	64.87	53.45	65.88
2	75.49	85.01	74.16	83.18
3	87.73	94.35	86.64	93.29
4	95.95	96.59	94.55	98.71
5	98.67	99.97	98.36	99.02
6	99.76	100.00	99.74	99.93
7	99.98	100.00	99.92	99.97
8	100.00	100.00	100.00	100.00

in the table the probability of classification differing from the correct one by at most one (the row with $radius = 1$) exceeded 0.52 in the case of 1H classifiers. Better results were obtained for 2H classifiers, where the respective probability exceeded 0.64.

Application of the interleaved training led to slight improvement in the ratio of perfect predictions (the row labeled $radius = 0$) and its much lower standard deviation (the limit of 100% was accomplished already for $radius = 6$), however the improvement was overall much less significant than in the case of HDR problem. The above preliminary results accomplished by the 2H classifiers are promising, but certainly below the level expected in professional applications.

Several possible paths can be explored in future research. The most natural one is investigation of the influence of the choice of square S on the quality of results - intuitively choosing S in the center of the board should be more beneficial than the currently tested choice of square $g5$. Another prospective direction is combining several classifiers built based on various squares in a mixture of experts approach with appropriate weighting scheme.

Move's Influence

The other method of move ranking in chess discussed in this chapter is proposed by Greer [143]. The method relies on pattern-oriented classification of moves based on heuristically defined *influence* of a particular move on certain board regions. At first, each square is assigned a label that represents heuristic belief in which of the playing sides controls this square. Combining this information for all squares leads to the *chessmap* [144], which represents the regions of the board that are under influence of either of the players, as well as the neutral areas, where none of the players has an advantage[3]. The chessmap's value of each square is calculated based on the following premises: whether the square is occupied or empty; in the former case whether it is attacked or not; if attacked the value depends on the ultimate result of the potential capture sequence on that square - the winning side controls the square; if not attacked but defended the control belongs to the side occupying the square. If the square is empty the side, which can move higher valued piece on that square controls the square. In the remaining cases (the square is occupied but neither attacked nor defended, or the square is empty and the highest ranks of pieces that can move on that square are equal for both sides, including the case that none of the sides can play a move on that square) neither of the sides controls the square and associated chessmap's value is neutral, i.e. equal to zero. If white controls the square the respective chessmap's value is equal to $+1$, if it is black - the value equals -1.

Additionally, for each square (or more generally each sector composed of some number of squares) the so-called *valueboard* is defined as a sum of values of all pieces that attack this square (either directly or indirectly) and a value of a piece occupying the square.

Finally, the *influence* of a move on a given square (or sector) is defined as the sign of the difference between the valueboard after that move and before (i.e., in a current position). This allows detection of squares that would be strengthened by that move, as well as the ones that would be weakened, and the ones not affected by the move.

In order to learn the influence relationship for chess positions a neural network was trained based on 10 000 positions extracted from master and grandmaster games. The input layer consisted of 70 neurons representing chessmap values of 64 squares and the kings' locations (the remaining six inputs). Each king's location was defined as being either on the queenside or on the kingside or in the center, and the respective neuron was set to $+1$ whereas the remaining two were equal to -1. The desired output values in the 64-element output layer (one neuron per square) were the influence labels after a move had been made in the actual game. After training, the outputs of the network were used to order board squares according to their predicted influence values. Consequently moves that influenced the highest ranked sector(s) of the board were considered as the most promising ones.

The above procedure was further enhanced by giving priority to forced and capture moves[4] in the following order: first the safe capture moves, then safe forced moves, then safe other moves, followed by unsafe capture, unsafe

---

[3] The idea of calculating the *influence* of white and black pieces in order to divide the board into sections controlled by the respective players was initially introduced by Zobrist in Go [353].

[4] A forced move is defined in [143] as one, in which a chess piece is forced to move in order to avoid being captured with resulting lost of material.

forced, and unsafe other moves, where unsafe move, as opposed to safe one, is the move which causes loss of material on the square that the piece is moving to.

Several tests of this interesting pattern-based heuristic were carried out including the set of 24 Bratko-Kopec positions [173], for which the quality of the method - calculated as the number of searched nodes - was comparable to the result of applying the history heuristic. The best overall result was obtained when the chessmap heuristic was supported by another heuristic that at each depth stored the last sector (square) that caused a cut-off. This sector was considered first at the related depth regardless of the neural network output.

Pattern-based pre-ordering of moves is in line with psychological observations of how human grandmasters make decisions about which moves to consider first. As Johannes Fürnkranz stated (referring to deGroot's studies [78]) in his excellent review of a decade of research in AI and computer chess [120]: "the differences in playing strengths between experts and novices are not so much due to differences in the ability to calculate long moves sequences, but to which moves they start to calculate. For this preselection of moves chess players make use of chess patterns and accompanying promising moves and plans."

A discussion on search-free playing methods is continued - in the spirit of the above quotation - in chapter 12 devoted to *intuitive game playing*.

Modeling the Opponent and Handling the Uncertainty

Modeling the opponent's playing style is a fundamental issue in game research. In general, the problem of modeling other agent's behavior extends far beyond game playing domain and is considered a crucial aspect in any competitive multi-agent environment, for example in decision support systems, stock market investment, or trading systems.

In game playing, the objective of modeling the opponent is to predict his future actions. Furthermore, by recognizing the opponent's way of play it is possible to exploit its weaknesses or lead the game to positions less convenient for the opponent. The potential gain strongly depends on particular type of game. Lesser impact is generally achievable in perfect-information games, in which all game related data is accessible, and where it is, in principle, sufficient to make the *objectively* best move at each encountered position in order to achieve the highest possible score. Definitely, no one can do better than that (assuming perfect play of the opponent).

In practice, however, the above strategy is hard to be followed if only the game is sufficiently complicated. Instead, the AI/CI player usually makes the *subjectively* best possible move according to its evaluation function and given time constraints. Consequently, also in perfect-information games the problem of modeling the opponent is not negligible. The style of opponent's play (tactical vs. positional, aggressive vs. defensive, etc.), if properly modeled, can provide an important indication for a game playing program. For instance, in a disadvantageous position a program could use a specific style of playing in order to hinder the potential victory of the opponent and strive to achieve a draw. Another example is seeking the chance to win an even game by steering it to "inconvenient" (for the opponent) positions and thus provoking the opponent's mistake.

The relevance of the opponent's playing style can easily be observed among human players. Practically each top player in any popular board game can point out the opponents who are "inconvenient for him", i.e., achieve relatively better results against that player than is indicated by their official

J. Mańdziuk: Knowledge-Free and Learning-Based Methods, SCI 276, pp. 169–180.
springerlink.com   © Springer-Verlag Berlin Heidelberg 2010

ranking. This biased score is most often caused by the specific style of play-ing represented by that opponent, which for some reasons is inconvenient for the player. In the above sense the "winning relation" is nontransitive and the ranking points provide only statistical estimation of the player's strength, rather than its absolute assessment.

Modeling the opponent's way of play is far more important in imperfect-information games. Unlike in perfect-information games, where the sole goal is to discover the weakness of the opponent's play by means of seeking the types of game positions unfavorable to him, the primary goal of modeling the opponent in imperfect-information games is related to revealing the hidden data. Efficient estimation of hidden information: the yet uncovered cards, tiles or dice, either possessed by the opponent or stored in a common game storage (cards pile, tiles bag, etc.) is usually the key issue in imperfect-information games.

In games of this type, the player is also faced with the "reversed" task, i.e. the problem of his own *unpredictability*. The most popular and generally applicable method of handling this problem is introduction of some amount of noise into decision making process, which disturbs the move selection policy, and thus hinders tracing the reasons of playing decisions and uncovering the playing style.

In the remainder of this chapter the problem of modeling the opponent is discussed from the perspective of five particular games. Four of them are imperfect-information games: RoShamBo, Perudo, poker, Scrabble, and the fifth one is chess.

RoShamBo

A simple though instructive illustration of opponent modeling problem is the kids' game Rock-Paper-Scissors, also known as RoShamBo [29]. In this game, each of the players independently and simultaneously chooses among Rock, Paper and Scissors. In the simplest case of two players the winner is the one whose choice "beats" the opponent's choice under the following rules: Rock beats Scissors, Scissors beat Paper and Paper beats Rock. If both players choose the same object the turn ends with a draw.

Even though the rules of the game are trivial, the game itself is more demanding that one might expect at first glance. A simple solution is of course choosing actions randomly according to uniform distribution. Such approach would statistically lead to a draw - one third of the games would be won, one third lost and one third of them drawn. This policy, however, does not take into account the opponent's way of play, which may possibly be inferred from his previous actions. In a trivial example, if the opponent always chooses Rock, the optimal policy is to choose Paper all the time, whereas the uniform choice policy described above would still statistically lead to a draw. Moreover, apart from trying to predict the opponent's next move, a skilful player should also avoid being predictable himself. Hence, simple rule-based approaches are not sufficient in this game and more sophisticated methods need to be employed.

Reader interested in development of adaptable RoShamBo playing agents may find useful the Internet reports [30] from the First and Second International RoSamBo Programming Competitions organized by Darse Billings in 1999 and 2000, respectively.

Perudo (Liar's Dice)

Another example of a game in which opponent modeling is critical for efficient playing is a dice game Perudo [200], also known as Liar's Dice. The rules of Perudo are not very complicated, though not as simple as those of RoShamBo.

The game is usually played by between 2 and 6 players and each of them starts the game equipped with 5 dice. The following excerpt from [191] describes the basic set of rules of the game: "All players roll and check their dice in secret. Then players take turns to bid on the total pool of dice. Each bid is an estimate, guess, or bluff, that there will be at least N many dice of a given value face, D. Following each bid, the next player must either make a higher bid or call. After a call all hands are revealed. If there are less than N dice of face value D then the bidding player loses, and has one dice removed. Else, the calling player loses, and has one dice removed. Any dice with a face value of 1 are considered wild - and count towards the total number of dice whatever D is. To make a new bid, the new bidding player must either increase D while keeping N constant or increase the number, N, or raise both D and N. Thus, bidding can be raised from three 3's to four 2's or three 4's or four 4's, and so on."

There are several variants of Perudo. The one described above represents the most popular version, which is considered in this chapter. For more details and discussion on possible game modifications please refer to [201].

One of the basic research questions concerns the ability of AI/CI player to guess (model) the opponent's hand based on analysis of his past bids. The problem is already demanding in a two-player variant of the game, and its complexity further increases in a multi-player case.

Playing the game well requires quite sophisticated analysis of opponents' past actions in order to detect possible bluffing, since the game significantly consists in bluffing and straightforward playing would most probably not lead to success against experienced players. Also, as described in [191], simple heuristic relying on probability distributions of possible dice outcomes can be easily outplayed by moderately strong human player. Generally speaking, in a two-player case any deterministic heuristic is likely to be reverse engineered by an experienced opponent, who should be able to infer the probable D and possibly the number of dice with face D, held by the player.

Perudo, especially in its multi-player version, is an interesting test bed for AI/CI opponent modeling methods [191, 200]. It is, to some extent, similar to poker in a sense of hiding one's dice/cards and making bids in turns. On the other hand, Perudo is much simpler than poker and, unlike in poker, at the end of each game the whole pool of dice is revealed (this happens in poker only under certain conditions). This way each game brings new data characterizing the opponents' behavior. The other important distinction in comparison with poker is that in Perudo there are no cash bids, which makes the bidding *directly linked* to the goal of the game - in poker most of the bidding phase in focused on cash and only *indirectly linked* to cards possessed by the players [191].

Perudo is relatively less explored than other games, while at the same time sufficiently complex to require more sophisticated approaches to modeling the opponents and covering player's own style of play than simple statistical estimation of probability distributions based on past decisions of game participants. The issue of modeling the opponent's behavior in Perudo is still almost uncharted territory.

Poker

Poker is undoubtedly the most popular "game of bluffing". Here, the notion of optimal playing is hard to define due to a huge amount of uncertainty regarding hidden opponents' cards (the *hole cards*) and the latent *community cards*[1]. Hence the optimal behavior strongly depends on the opponents' actions and can only be estimated with some probability. A simple and instructive example concerning the frequency of bluffing by the opponent is given in [31]. Generally speaking, it is advisable that the player who bluffs more frequently be called more often compared to the one who bluffs relatively rarely. In other words, the frequency of calling should be adjusted to the opponent's individual bluffing pattern.

The problem of modeling the opponent becomes much more demanding in the case of multi-player game - which is actually a typical situation - when, due to increasing amount of uncertainty and various possible strategies implemented by the players, the context of decisions made by each game participant is much more complex than in the case of two-player game.

On a general note, two strategies can be employed to predict the opponent's next move. The first one relies on applying our own betting strategy, possibly with some minor refinements. This approach leads to a model which is of general applicability and outputs "reasonable" decisions adequate to our beliefs and experience in the game, but makes no use of specific information about the style of play presented by a particular player. Considering this player-specific information seems, on the other hand, to be indispensable in achieving high performance in poker.

The other possibility of modeling opponent's future behavior is calculation of the probabilities of possible opponent's next actions based on analysis of sufficiently large number of games already played against that opponent. This

---

[1] We refer to the Texas Hold'em variant of poker, which is the most popular version of this game - see the footnote in chapter 3.3.

approach assumes that the opponent will continue to behave in a way he used to do in the past.

The heart of such a model is an estimator of the likely probability distribution of the opponent's hand (two hidden hole cards). Since there are 1326 possible two-card hands one of the convenient ways to handle these probabilities is maintaining a *weight table*, in which the estimated probabilities are scaled between 0.1 and 1.0 [77]. In the simplest case, when no opponent modeling takes place, the values in the table are fixed and not modified any further. Otherwise, after each player's action the respective values are recalculated in order to be consistent with the currently modeled betting strategy, leading to re-weighting the estimated probabilities. Probability distribution may be estimated basing on some general assumptions enhanced by the analysis of the betting history and statistics of each opponent [31, 77]. These general factors which may be accounted for refer to the objective issues such as the advancement of the game, the number of players still in the game, or the number of those that have folded and those that have called. For example, after the flop, the probability that a player holds an *Ace* and a *Queen* (i.e. relatively strong cards) is higher than that he has e.g. *eight* and *two*, since most of the players having *eight* and *two* would fold before the flop.

Another possibility, besides using a weight table, is to train a neural network to predict the opponent's next action. The Poki-X system (introduced in chapter 4.5) developed at the University of Alberta employed a standard, one-hidden-layer MLP with 19 inputs representing particular aspects of the game, 4 hidden units, and 3 outputs corresponding to the three possible opponent's decisions (*fold, raise,* or *call*). The outcome of the network, after normalization, provided a probability distribution of the possible opponent's actions in a given context defined by the input features. After training on hands played by a given opponent, magnitudes of network's weights reflected the relative impact of particular input features on the predicted outcome, which allowed further exploration of the available data.

Initial results of weight analysis revealed two particularly relevant features strongly influencing the next opponent's action: the last opponent's action and the previous amount to call [77]. In particular, a strong correlation between the predicted decision and the last action was observed in case the last opponent's action was *raise*. Detailed analysis of network's connections resulted in defining a relatively small number of context-sensitive equivalence classes in the input space, which significantly improved the results of prediction [31].

Another issue of great importance in poker is the problem of *unpredictability*. As mentioned at the beginning of this chapter, one of the most popular approaches to this task is occasional changing of our betting strategy, even at the cost of its local quality deterioration. Introduction of such random divergences from the main strategy seriously hinders other players' attempts to accurately model our way of play and usually leads to incorrect beliefs about our betting policy, consequently provoking opponents' mistakes. One of the possible realizations of the above idea is proposed in Poki-X bot, in which after estimating the normalized probabilities of each of the three possible actions (*raise, call, fold*), a final decision is made based on a randomly sampled number, with uniform probability, from the interval $[0, 1]$.

The task of opponent modeling is one of the most challenging issues in poker. From the psychological point of view, the problem may not be perfectly solvable in practice since, as stated in [31] "even the opponents may not understand what drives their actions!". Nevertheless there is still a long way until artificial agents approach the level of top human players' abilities in this area.

Simplified Poker

As discussed above the two major sources of uncertainty in poker are related to the unknown cards' distribution (cards possessed by the opponents as well as the latent community cards) and the unknown opponents' playing styles. The latter factor is intrinsically dependant on the opponents' individual traits, such as preference for aggressive or conservative play, the degree to which they accept risk in the game, as well as several momentary factors such as current mood or physical disposition. Consequently, for any particular cards' distribution the course of the betting phase may significantly vary depending on the particular choice of players. Moreover, it can also vary even for the same configuration of players, but at another time.

Considering the above, some researchers attempted to minimize the overall amount of uncertainty by strongly reducing the diversity of potential hands in poker, thus focusing exclusively on modeling particular styles of play. The natural path to performing such studies is taking into account simplified versions of the game played with a subset of the entire deal and with lower number of cards per hand. Typically, one-card versions of poker [52, 172] are considered in this context, in which a deck is composed of fewer cards than in full-scale poker and each game participant is given one card only.

A very recent example of such approach was presented by Baker and Cowling [15], where the authors considered a four-player game, with a deck composed of 10 pairwise different cards, ordered according to their strength with no concept of suit. In each game, each of the players is dealt one card and then the betting takes place with the cost of each bet fixed at one chip. The maximum number of possible betting rounds is limited only by the number of chips possessed by the players. Actions possible at each turn are *raise, check, fold*, i.e. the same as in regular poker. Once the betting phase is completed the player with the highest card among those who remained in the game becomes the winner and receives the pot contents. Initially, each player is given 10 chips. The game ends when one of the players collects all chips.

Following [17] the authors defined four major categories of poker opponents: Loose Aggressive (LA), Loose Passive (LP), Tight Aggressive (TA) and Tight Passive (TP). The loose players are generally those who overestimate the strength of their hand as opposed to tight players, who are quite conservative in this respect. Aggressive players are the ones who strive to increase the amount in the pot, whereas passive players are rather reluctant to raise the pot. These four styles are commonly parameterized by two factors: α and β, where α denotes the minimum win probability in order for the player to stay in a game (not to fold), and β represents the minimum win probability for the player to bet. Certainly, the following conditions are true:

$$\alpha(Loose) \leq \alpha(Tight), \qquad \beta(Aggressive) \leq \beta(Passive) \qquad (11.1)$$

where $\alpha(\cdot), \beta(\cdot)$ denote the values of respective parameters for particular types of players. Additionally, $\alpha(X) \leq \beta(X)$ for any $X \in \{LA, LP, TA, TP\}$ by definition of α and β.

(α, β) pairs describing each of the four types of players were chosen arbitrarily. Next, each player was confronted with its "anti-player" in a systematic test of several values of α and β assigned to anti-player within the constraints $0 \leq \alpha \leq \beta \leq 1$. In each testing 100-game match the anti-player was pitted against three instances of the respective player. The best (α, β) pairs found in this anti-player selection process (one for each of the four types of players) achieved scores between 63% and 76%, depending on the player's type.

Furthermore, the authors applied Bayes' theorem in order to define the universal anti-player, i.e. one which is a worthy opponent regardless of the nature (type) of the remaining players. First of all, the conditional probabilities of the opponents' playing styles on condition of particular undertaken action (raise, check, fold) were calculated by analyzing thousands of games played by the respective opponent players (LA, LP, TA, TP). Next, a simple algorithm for choosing the appropriate tactic for universal anti-player was proposed, which consisted in recognizing each opponent's style (with the help of the above conditional probabilities) and depended on the opponents' demographics (the numbers of opponents of various types among the three contestants other than the anti-player).

A series of 100-game test matches was played by the proposed universal anti-player and various combinations of the remaining three players. The results achieved by the anti-player varied between 40% and 90% depending on the types of the players it was confronted with in a given match. The weakest outcome was obtained in the case the opponent players were (LP, LP, TP) and the highest one for the triple (TA,TP,TP). The results proved a general applicability of the universal anti-player against the four types of players defined in the paper in a four-player simplified poker game. Even the weakest result exceeded the "expected average" number of wins by 15%.

Another comparison was made with the results attained by the so-called simulation player, which knew the opponents' playing styles in advance and adjusted its parameters α and β to the actual demographic of the opponents by systematically testing a set of 66 (α, β) pairs in increments of 0.1 (for each parameter) with constraints $0 \leq \alpha \leq \beta \leq 1$. For a given set of opponents, for each tested choice of (α, β) the simulation player ran a simulation of a 100-game match against these opponents and chose the most effective α and

β for tournament play. The universal anti-player (which did not know *a priori* the types of opponents it was pitted against) achieved comparable results to those of the simulation player.

The universal anti-player proved to be effective against the fixed types of players. An interesting future comparison might be pitting the universal anti-player against diverse opponents, reflecting the whole spectrum of possible playing styles, i.e. the opponents which select their α and β parameters freely, not being restricted to particular preestablished selections. Furthermore, the applicability of the proposed approach to the case of opponents able to change their playing styles during the match is another interesting open question.

Scrabble

Scrabble, as mentioned in chapter 4.4, has been for years one of the main AI/CI targets in game domain. The best Scrabble programs: Maven and Quackle already have an edge over top human players. Due to stochastic nature of the game and the existence of hidden data - the private tiles on the opponent's rack and the tiles left in the bag - the game is an ideal test bed for implementation of methods that focus on modeling the uncertainty related to imperfect-information. Note, however, that even if the opponent's rack were modeled with perfect accuracy the game would not become a perfect-information one, since the tiles left in the bag would still be selected randomly introducing the element of luck into the game.

The question of how much one could have gained from knowing the contents of the opponent's rack was posted in [266], where the authors consider the problem of opponent modeling in Scrabble. Using the Quackle's Strong Player (QSP) program, a number of games was played between the QSP and a copy of the QSP having full knowledge of the opponent's letters (QSPFN). In 127 games the QSPFN scored on average 37 more points per game than the QSP, which in Scrabble reflects a great advantage. A standard deviation of the results was relatively high: the scores ranged between about 140 points in favor of QSP up to about 300 point victories of QSPFN. Such distribution is an expected phenomenon since even access to the complete knowledge of the opponent's rack and championship level of play do not guarantee the success in a particular game as it strongly depends on luck in sampling the tiles from the bag.

The above experiment confirmed the hypothesis that even in the case of close-to-perfection Scrabble players, as QSP is, there is still room for improvement when the opponent's tiles on his private rack are faultlessly predicted. The idea of how to approach this prediction task proposed in [266] relies on making inferences about the rack's contents based on the conclusions drawn from previous moves. In short, the modeling system assumes that the opponent is playing basing on perfect (i.e. complete) dictionary - which is generally a valid assumption in the case of computer players - and therefore, if in previous moves some possibilities of good scorings have been "missed", this must have been caused by the lack of certain letters at the opponent's disposal.

The model of the opponent is quite simple in that it expects the opponent's decision-making system not to involve any simulations. In other words, it is assumed that at his turn the modeled player makes the same move as we would have made being at his position, based on the same static evaluation function that we use. This is certainly a simplistic assumption since in real play simulations of depth 2-4 ply usually take place. Approximation of the opponent's decision process relying on his own simulations appeared to be too costly and therefore only static evaluation function was used in assessment of candidate opponent's moves.

The inference model of the opponent's rack was implemented based on the Bayes' theorem:

$$P(leave|play) = \frac{P(play|leave) \; P(leave)}{P(play)}, \tag{11.2}$$

where $P(leave)$ is a prior probability of particular leave (rack's contents after the move was made) - calculated as the probability of drawing a certain combination of letters from all letters unseen by the player. The $P(play)$ probability is calculated as:

$$P(play) = \sum_{leave} P(play|leave)P(leave), \tag{11.3}$$

$P(play|leave)$ denotes our model of the opponent's decision-making system. In the simplified assumption about always making the best move, if the highest-ranked word found in a given game position was the one actually played by the opponent then $P(play|leave)$ was set to 1, otherwise is was set to 0. This way

$$P(leave|play) = \frac{P(leave)}{\sum_{leave \in M} P(leave)} \tag{11.4}$$

where M is a set of all leaves for which $P(play|leave) = 1$.

Based on the above described model, at each opponent's turn, each of the possible leaves that the opponent might have had was combined with the set of letters played at that move in order to reconstruct the possible full rack's contents. For each such reconstructed rack the set of all possible moves was generated and that information was in turn used to estimate the likelihood of a given leave, assuming the optimal play in the sense that no "human-type" overlook errors took place and that a complete dictionary of legal words was available.

The above described inference procedure was implemented in the Quackle Strong Player With Inferences (QSPWI) version of the basic QSP. The 630-game match between these two programs resulted in the score of 324 : 306 wins in favor of QSPWI, with the average 5.2-point advantage per game. The difference is statistically significant with $p < 0.045$. The five-point average advantage is, according to the authors, a significant improvement in real play, which proves the practical value of proposed method of modeling the opponent's rack.

Certainly, exact approximation of the contents of the opponent's rack is critical for achieving a close-to-optimal play. It allows, for example, to play in a more risky and more aggressive way, when it is known that the opponent is unable to respond with a high-valued countermove. Additionally, it prevents occurrence of positions allowing the opponent to play a *bingo* (a move in which the opponent uses all seven of his letters in a single move - for such a play a 50-point bonus is granted). Similarly, knowing the opponent's hidden letters allows blocking the lucrative opponent's moves by playing on the relevant spots.

Chess

As it was mentioned in the introduction to this chapter the problem of modeling the opponent is less relevant in the case of perfect-information games. However, also for this type of games the role of modeling the opponent's *playing style* is not negligible. One of very few published attempts focusing on formalization of the intuitive concept of playing style in perfect-information games is presented by Arbiser [9], for chess. The author proposes a method of implementing several fundamental game-related terms such as attack, threat, sacrifice, etc. as well as the notions of aggressive, defensive, strategic, tactical, and positional play.

The underlying idea is based on generalization of the null-move heuristic [23, 148] introduced in section 3.2.3, in such a way that instead of (virtually) moving twice in a row, the opponent is allowed to (virtually) change one of his or player's pieces or add/delete a piece and then make a move.

For example, the notion of opponent's aggressive play may be implemented by changing one of opponent's or player's pieces into a strong opponent's piece before deciding a move. Such an exchange would most probably cause immediate threats to the player thus forcing the choice of an appropriate response.

A strategic move (play) of the player may be implemented according to the following scheme: move a piece and then let the opponent move two or more times. Certainly, if the move was in fact strategic and a game position allowed making such a long-horizon move, no immediate threat/loss in the player's position should appear within the next two (or more) moves. Otherwise, it is time to make a tactical or at least short-horizon move instead of a strategic one.

In brief the following scheme is proposed: modify the board in an adequate manner before calling a regular search algorithm and ensure that the search-based chosen move would be valid and sound in the original board position, i.e. without initial, fictitious modification.

Despite several doubts concerning the time complexity of the method and missing implementation details, the idea presented in [9] seems to have potential and is worth further consideration. After necessary modifications and refinements a method built on Arbiser's idea may possibly be utilized as a supportive tool in a chess playing program, in conjunction with other, more established methods.

Another approach to the opponent modeling in chess is presented in [334, 337]. The method is based on the psychologically plausible idea of *chunks of information*[2]. The proposed Inductive Adversary Modeler (IAM) system attempts to acquire the set of patterns representing typical opponent's behavior. On its turn the IAM searches the set of chunks in order to find the ones that can be completed in one opponent's move, assuming that having such an opportunity the opponent will willingly "complete" the chunk. In the case when there is more than one choice, the selection is made heuristically with preference for bigger chunks.

These likely opponent's moves, when used to prune the search tree, allowed for about 12.5% increase of the search depth in the tests on grandmaster games within a given amount of time [336]. Furthermore, a specially designed IAM was successfully applied to the problem of predicting grandmaster moves in the opening phase [335] with about 65% accuracy.

Summary

Computational Intelligence is very well suited and widely applied to the problem of opponent modeling. Probabilistic methods allow building either a generic opponent model for a given game, or specific model of individual player's behavior. An alternative approach, especially effective in case the opponent's patterns of activity vary in time, is to use neural networks, due to their capability of adapting internal parameters to the gradually changing input data. Another promising avenue is application of TD-learning to adaptation of internal parameters of the playing system in order to strengthen it against particular opponent.

Online, adaptable and close to reality modeling of the opponent is one of the fundamental challenging problems in any nontrivial game, in particular in imperfect-information games, in which the data hidden by the opponent can only be inferred by analysis of his past actions. Proper identification of this concealed information greatly increases the expected player's outcome.

The problem of opponent modeling becomes much more demanding when multi-player situation is considered, in which the opponents' decisions depend additionally on the number of contestants, the order of their move-making turns, and their previous bids in the current game. Moreover, in multi-player games the contestants can form *ad-hoc* or long-term coalitions (formal or informal) against the current leader or in order to gain some benefits. In such a case players' decisions are highly contextual and partly depend on the current objectives of the coalitions they belong to.

---

[2] The concept of suitable manipulation of appropriately defined *chunks of information* plays a leading role in psychological theories concerning game representation in human brains. These issues are discussed in detail in chapter 12 devoted to intuitive game playing.

Opponent modeling becomes also more and more important in commercial computer games, where one of the challenging goals is adaptation of the playing strength of artificial characters to the increasing skill level of a human player [122]. Such adaptation process is one of the aspects of a more general concept of *believability*, that is building AI/CI agents that are functionally indistinguishable from humans. Since commercial games are out of the scope of this book we only mention here that in the case of skill games, for example, the rivalry with computers can only make sense (and fun!) if machines play in a way similar to humans (i.e. occasionally make mistakes and are able to imitate the process of learning by gradually increasing their playing abilities according to human player's progress in the game). Otherwise, playing skill games with computers is like playing against unbeatable opponent.

Part IV

Grand Challenges

Intuition

Implementation of the concept of intuition is definitely one of the greatest challenges in computer games and also in computer science in general. Nowadays, despite the major breakthroughs made in several disciplines and increasingly deeper scientific understanding of the nature, intuition - paradoxically - becomes more important than ever. Regardless of particular problem domain the intuitive solutions applied by humans (or machines) have the two following common characteristics: they are reached rapidly (almost instantaneously) and it is impossible to explain analytically the way the solution has been invented.

In most of mind board games intuition plays a leading role at master level of play. Consider for example chess. With a branching factor of about 30, in a 100-ply game there are about 10^{147} contingencies, which is an enormous number for any human being (grandmasters are believed to search no more than a few hundred contingencies during the entire game). How then is it possible that chess champions are able to play at such a high level? One of the main contributing factors is intuition which allows them to perform highly selective search in this huge space. Another illustration of intuitive playing are simultaneous games. Chess grandmasters usually need only a few seconds to make a move, which generally proves to be very strong (often optimal). This means that they possess the ability to immediately find the most relevant information characterizing board position and recognize the most promising move, usually *without deep, precise calculation of all its consequences.*

The ability to almost instantaneously recognize the strengths and weaknesses of a given position allows grandmaster players, when they approach an unknown game position, to assess which side is in the winning or favorable situation within a few second time, in most of the cases. One of the possible psychological explanations of this phenomenon (discussed below in this chapter) is the ability of advanced players to link the new position with previously explored familiar ones and consequently focus on moves and plans associated with these, already known, positions [78].

In the above context, one may provide an operational definition of intuition as an instantaneous, subconscious recognition/reasoning process which does

J. Mańdziuk: Knowledge-Free and Learning-Based Methods, SCI 276, pp. 183–204.
springerlink.com

not rely on precise, deep calculations, but instead rather refers to past experiences and previously acquired general knowledge. Consequently, in most mind games, intuition is one of the main factors contributing to the beauty and attraction of the game. Its application may lead, for example, to long-term material sacrifices without apparent possibility of its recovery.

A well-known illustration of intuitive playing in chess is the *Immortal game* played in London in 1851 by Adolf Anderssen and Lionel Kieseritzky, in which white sacrificed bishop (on move 11 - see fig. 12.1(a)) and subsequently two rooks and a queen (starting on move 18 - see fig. 12.1(b)) in order to checkmate on move 23 - (fig. 12.1(c)). Certainly, the last three sacrifices were tactical ones, i.e. their consequences could have been precisely calculated by Anderssen, but the introductory sacrifice (bishop on move 11) is an example of intuitive playing based on player's experience and "feeling" of the board position.

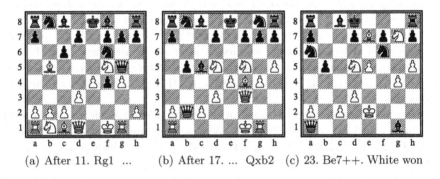

(a) After 11. Rg1 ...   (b) After 17. ... Qxb2   (c) 23. Be7++. White won

Fig. 12.1. Anderssen vs. Kieseritzky, *Immortal game*, London, 1851

Another telling example of intuitive sacrifice occurred in the game played between two great archenemies: Anatoly Karpov and Garry Kasparov in the New York match in 1990. In the middle-game position Kasparov sacrificed queen for a rook and knight on moves 16 − 17 (see fig. 12.2) and this sacrifice was clearly positional with no immediate tactical or material threats. The game continued up to 53th move, when players agreed to a draw.

An instructive example of an intuitive position was presented by Wilkins [346] and also considered by Linhares [190] in the presentation of his Copycat-like chess playing system (described in chapter 12.4 below). In the position presented in fig. 12.3 a few quite obvious observations are easy to make even by a moderately experienced chess player. First of all, pawn structures on both sides are mutually blocking each other and the only pawn "free to go" is white piece on f6. The possible advancement of unblocked white pawn restricts the area of black king movements, since the king must hinder possible promotion of that pawn on f8. Finally, there is no possibility for any piece

(a) After 16. Nb6 ...

(b) After 16. ... axb6, 17. Rxd7 Bxd7

Fig. 12.2. Karpov vs. Kasparov, New York, 1990

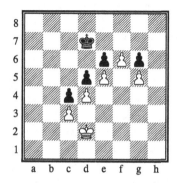

Fig. 12.3. An example of position, after [190, 346], in which intermediate human chess player would easily find the game plan without any search, by noticing that (1) all pawns are blocking each other, except for white pawn on square f6, (2) black king is simultaneously guarding black pawn on e6 and preventing white pawn's promotion on square f8 and therefore is tied to this area, (3) white king can approach black king on the left (queen) side of the board and force him to give up one of his two tasks

(kings in particular) to reach the other side of the board by moving on the king-side (lines e-h), but there is such a possibility on the queen-side (lines a-d) using the queen rook line.

Considering the above it is relatively straightforward that the winning plan for white is moving their king to the other side of the board via the leftmost column and forcing the black king to either abandon the defence of black pawn on e6 (still trying to control square f8) or to actively confront the white king (moving towards him), but this way letting promotion of white pawn f6. Such a plan, being a result of rather simple analysis, is quite obvious to human players (playing at above-novice level) and intuitively achievable practically with no search at all. On the other hand, typical computer-playing approach, would require search as deep as 19 ply in order to virtually reach the position of white king pressing the black king to abandon guarding squares e6/f8 [346].

Certainly, one may argue that 19-ply search is achievable for machines (and there is no doubt about it - *vide* Deep Blue's search capacity), but the question is actually the following: what would happen if a similar idea was implemented on 50×50, or 100×100, or yet greater board? Human players would, in principle, use the same type of analysis and reach qualitatively the same conclusions. How about computers?

12.1 Perception in Chess

Some of the most salient research studies focused on understanding the concept of intuition in game playing in the context of chess were performed by de Groot [78, 79] and Chase and Simon [60, 61]. In de Groot's experiment chess players (of various skill levels) were shown chess positions, each having about 25 pieces for a short period of time (between 2 and 15 seconds) and then, after removing the position from a view, asked to reproduce it. The accuracy of reconstruction among grandmasters equalled about 93% and systematically decreased as a function of decreasing skill.

In a similar experiment Chase and Simon [60, 61] confirmed independently that given a five second time of exposure of a chess position the quality of reconstruction among chess masters was visibly higher than that among less advanced players. Moreover, in case of random positions the difference between players of different skills was insignificant, which strongly suggests that the advantage of more skillful players should not be attributed to their better memory functioning, but rather to the fact that their perceptual encoding of a board position reflects deep understanding of subtleties of chess. Chess masters are believed to use recognizable chunks (templates) of a given position as indexes to the long-term memory structures which in turn trigger the use of plausible moves in the search process. Such a protocol allows significant reduction of the set of moves to be considered. The size of expert's "chunks library" representing specific combinations of pieces was estimated at approximately 50 000 in early studies [301] and around 300 000 in more recent publications [139].

Some of the postulates concerning functioning of intuitive processes in human masters' brains are supported by experimental evidence from saccadic eye movement tracing. De Groot and Gobet [80] demonstrated that skilled players made more fixations *along the edges* of the board squares than the less advanced players who are more focused *on squares*. Moreover, the average distance between the two successive fixations of master players is greater than that of intermediate or novice players. Both the above observations suggest that strong players encode two or more pieces in a single fixation rather than a single piece - as is the case of less experienced players.

Subsequent works [59, 263], aimed at investigation of spatial distribution of the first five eye fixations of the players attempting to choose the best move in a given position, were consistent with the findings of de Groot and Gobet, and indicated a greater distance between consecutive fixations and higher proportion of fixations on salient pieces in case of skilled players.

In order to further verify/support Chase and Simon's hypothesis about addressing chess masters' memory by chunks of a board position the authors of [263] performed a specially designed series of check detection experiments in which both expert and novice players were verifying the checking status on a small 3×3 or 5×5 chess boards. Each position was composed of a king and one or two potentially checking opponent pieces. In some of the experiments one of the two potential attackers was cued (colored). In another group of tests chess pieces were represented by letters (K, R, N, ...) instead of traditional symbols (see [263] for details).

Results of these experiments clearly indicated that chess skills are much more related to experience of players than to their general perceptual superiority. In particular, in the case of replacing chess pieces' representation by letter symbols, the number of fixations in a given time interval increased for both expert and intermediate players. This phenomenon was not observed for novice players. It was therefore concluded that there must exist a relation between the chess playing skills (experience in playing the game) and the specific internal representation of game positions manifested, in particular, by visibly greater familiarity with symbol notation than letter notation in case of experienced players.

Results attained in the case of having one vs. two potentially attacking pieces as well as results of experiments with coloring one of the pieces show that experienced chess players perform position analysis in an *automatic* and *parallel* manner - they do not benefit from cueing a piece or simplifying position (from two pieces to one) in terms of the average time required to answer the question whether or not the king is being checked. Hence, it can be concluded that one of the possible mechanisms underlying high playing competency may be "parallel extraction of several chess relations that together constitute a meaningful chunk" [263].

12.2 CHREST/CHUMP Model

One of very few implementations of human perception mechanisms in artificial chess-related systems was proposed in the CHUMP experiment [138], based on the Chase and Simon's chunk theory. CHUMP relies on the Chunk Hierarchy and REtrieval STructure (CHREST) computer model of human perception and memory in chess and directly implements the eye-movement simulator to scan the chessboard into 20 meaningful chunks.

In the learning phase a move which has been played in a given position is added to a specially designed discrimination net [137] and linked to the chunk containing the piece played and its initial location. In the test (play) phase the patterns retrieved in a given position are checked for the associated moves and the one which is proposed by the highest number of chunks is played.

The performance of CHUMP was tested on the set of widely used Bratko-Kopec benchmark positions [173], for which the system was rated below 1500 ELO (intermediate amateur player level). It should be noted, however, that due to the lack of search mechanisms the system was handicapped in tactical positions and therefore its ability of positional play should be rated relatively higher.

Another set of test was performed on a collection of games by Mikhail Tal and in the KQKR endings. In many positions CHUMP was capable to retrieve the correct move, although it usually wasn't the highest ranked one [121].

The main weakness of CHUMP is its inefficiency in maintaining the set of meaningful chunks, which grows linearly with the number of training positions. This suggests potential problems with reusing already possessed knowledge, since CHUMP acquires new chunks, instead. Despite the above drawback, CHUMP is an important example of a system operating without any search, purely driven by pattern recognition and pattern selection mechanisms. It was also the first chess program among those based on psychologically motivated human memory and perception models, capable of playing the entire chess game [120].

12.3   SYLPH: A Pattern-Based Learning System

In 1998 Finkelstein and Markovitch [106] proposed a pattern-based approach as a possible representation of human intuitive reasoning in chess. The system was in part related to the experiences with Morph - a famous chess learning system [185, 188] briefly introduced in chapter 13 - but the general inspiration of applying pattern-based move-oriented approach came from Gobet and Jansen's CHUMP.

In short, SYLPH relies on the so-called *move patterns*, each of which consists of a board pattern and a move associated with the game position represented by this pattern. Additionally, each move pattern has a weight assigned to it, which indicates a potential benefit of applying a given move in the context provided by the pattern. Move patterns are used to prune the search tree by selecting only the most promising branches for further exploration, thus narrowing the search space and extending the search depth given the same amount of time.

The language used to define move patterns is an extension of that of Morph. Board patterns are represented as graphs whose nodes are chess pieces or empty squares and edges represent relations between the nodes. Examples of the simplest relations include: *a piece controls an empty square* or *a piece directly attacks/defends another piece* or *a piece indirectly attacks/defends another piece through another piece*, etc. More complicated relations may involve four or five pieces or empty squares. The method of populating and updating pattern database is developed in a similar way as in Morph with extensions allowing for hierarchical storage of the patterns.

Patterns are learned in two modes: the main one is by playing with a teacher, which can be either a human or a stronger program (GNU Chess here) or a copy of itself. The other, auxiliary mode of learning is based on self-generated training examples, by simulating the unplayed, but potentially valuable moves (according to the learner). After simulating the unplayed move the pattern generation process proceeds in the same way as in the regular learning mode.

In either of the two modes, patterns are extracted and memorized based on the reversed trace of the game. Board patterns for which the evaluations *before* making a move and *after* the opponent's response are very different are used as training examples of the *material patterns*, since the difference in assessment is usually caused by a change in the material balance. Another type of learning patterns, are *positional patterns*, which are assumed to be meaningful when the moves associated to them were performed by a strong trainer (human or artificial). These patterns/moves do not lead to changes in material balance and are utilized, for instance, to learn the openings - see [106] for details.

In the case of material patterns weights assigned to them reflect the respective material difference. For positional patterns the weight is proportional to the frequency of using the pattern in the teacher's play. In order to make this weighting more statistically significant the so-called augmentation process is performed which consists in observation of several games played by two copies of the teacher or in analysis of the masters' games database in order to update the statistics and the corresponding weights.

SYLPH, which is the abbreviation for SYstem that Learns a Pattern Hierarchy, was trained by playing 100 games against GNU Chess, which was followed by an augmentation phase of observing 50 games played between two copies of the teacher. During training SYLPH acquired 4614 patterns. The efficacy of training was measured by system's ability to filter the best moves in unknown game positions extracted from games between two copies of GNU Chess - these games were not used in the augmentation phase.

The filtering task was defined in two ways: the filter $f_k, 1 \leq k \leq 10$ was selecting the top-k moves, and the filter $f_\gamma, 0.1 \leq \gamma \leq 1.0$ was selecting the γ fraction of the best moves. The success rate was measured as the number of test positions in which the best move was included in the filtered fraction of presumably best (according to SYLPH) moves. Various statistics and comparisons were reported in the paper. The most intriguing was the fact that after only 30 training games the f_γ filter with $\gamma = \frac{1}{3}$ outperformed a material-based filter that applied alpha-beta search with depth 4.

In order to test the generality of the method the system was tested on 3830 moves taken from 100 games played by Mikhail Tal (the former World Chess Champion). The curves representing the success rate versus γ or k were of very similar shape as in the case of tests performed on GNU Chess positions.

For comparison purposes also CHUMP system was tested on the same set of positions and yielded results inferior to SYLPH. For example, the success rate of SYLPH for $k = 4$ was equal to 0.577 compared to 0.194 rate of CHUMP. Finally, the alpha-beta search with $d = 4$ and evaluation function based on material values was pitted against the alpha-beta algorithm with $d = 4$ which applied the $f_k, k = 5$ filter in the root of the search tree. The selective f_5-based system managed to tie a 50-game match with regular alpha-beta with 6 wins, 6 losses and 38 draws, while searching smaller trees.

In summary, SYLPH performed unexpectedly well considering its short 100-game training regime starting with rudimentary domain knowledge only (game rules, material assessment of the pieces and the set of predefined basic relations). It is a pity that, to the author's knowledge, this work was not pursued any further. One of the possible impediments in further extension of SYLPH project, as suggested by Finkelstein and Markovitch, could be time complexity of the process of filter application as well as problems with managing growing patterns database. In order to efficiently handle an increasing number of move patterns, some retention mechanisms would have to be implemented allowing removal of patterns with low utility.

12.4   COPYCAT-Like Model

One of the most developed, well-grounded proposals of realization of human-type intuition in machine game playing was introduced by Linhares [190] with regard to chess, based on the theory of *active symbols, abstract roles* and multi-level *distance graphs between board pieces*.

The underlying concept of Linhares's model is, according to the author, its psychological plausibility and conformance to experimental evidence concerning the grandmasters' perception and memorization of chess positions, partly described above in this chapter. This plausibility is accomplished by fulfilling seven architectural principles defined in [190]. Apart from the principles related to chess position analysis performed by human expert players (focused search, limited number of moves to be analyzed, concentration on relevant pieces and relevant empty squares, formation of chunks in the short term memory (STM) in parallel mode, maintaining virtual chunk-library in the long term memory (LTM)), the playing system should also apply bottom-up and top-down processing modes in parallel in order to form meaningful board representations on various levels of detail.

In order to implement the above principles it is proposed in [190] to adopt the design principles of Copycat project [216, 226, 227]. Copycat is a cognitively plausible architecture invented to model human-type perception of sets of letter strings. The system is initially equipped with very basic knowledge consisting only of an ordered set of 26 letters of an alphabet. Copycat uses the STM as a working area for emergence of various concepts which compete with each other under several impulsive processes (called *codelets*). These processes are executed in parallel and each of them attempts to accomplish the goals and priorities assigned to it.

The LTM is constructed in the form of a network. The nodes represent the concepts (originally developed in the STM) and the weights denote relations between particular concepts. Activation of any specific concept triggers the chain reaction process propagated through the network in an unsupervised manner, spreading the activation to various nodes (other concepts). The LTM network is called *slipnet*, as it is able to "slip one concept into another" [149, 190, 226].

On the operational level Copycat uses a control parameter - *temperature*, which steers the competitive process of concept formation in the system. This is a close resemblance to the *simulated annealing* method, where the temperature drives the system's state (problem's solution) from initial high disorder to gradually more and more organized form, along with temperature decrease. Similarly to simulated annealing Copycat commences its operation at relatively high temperature allowing for temporal formation of different occasional concepts or sub-solutions which emerge autonomously, in an *ad hoc* manner, but are incapable of surviving in unchanged form for a long time. The lower the temperature the more reliable and more meaningful the emerging concepts. Gradually the chunks formed in the system's working memory compose the final system's output - a solution to the problem being solved. There is, however, a significant distinction between classical simulated annealing and Copycat system, since in the former the temperature is monotonically decreasing in time, whereas in Copycat the increase of temperature is possible and may happen quite frequently. This property stems from the fact that the temperature schedule in Copycat in not governed externally (as is the case of classical simulated annealing implementations with predefined temperature schedules), but is adapted internally by the system itself.

In general, the way the Copycat system operates in the domain of letters resembles to a large extent human intuitive processes [216, 226]. According to Linhares, the implementation of Copycat-like architectural and operational principles adapted to particular game-playing task is a feasible way to build a machine capable of intuitive playing akin to human way. In particular, with regard to chess-playing systems it is proposed in [190] to pay attention to the following three major aspects.

First of all, the key role of *active symbols* processing in Copycat's STM and LTM described above is emphasized. In short, active symbols represent parts of temporal solutions and emerge in the collective computation process driven by certain urges and partial goals. These symbols trigger chain reactions, spreading the information along various pathways in a quest for ultimate construction of the final solution.

The second underlying Copycat's concept adopted in proposed system is the notion of *abstract roles*. Each chunk has a set of data structures (relations) associated with it that intend to capture the roles it plays in a given position. For example, a single-piece chunk may be an attacker, a defender, a blocker, etc. Moreover, a chunk (piece) may perform its roles to a certain degree of (estimated) efficiency, e.g., a queen defending pawn is in most cases a weak defender, whereas pawn simultaneously attacking two opponent's pieces is a strong attacker, and so on. Similar abstract roles can be defined in terms of mobility or king's safety. Certainly, each chunk may play several different abstract roles simultaneously depending on its relations to other chunks (pieces) in a game position. It should be stressed that although the roles are assigned to chunks formed in a certain situation, their notion is general, since they *abstract* from specific location of a chunk, or distances, or trajectories in a position. As stated in [190], such abstraction leads to the emergence of intuitive strategies in which the reasoning is related to roles rather than exact locations.

A meaningful illustration of the idea of *abstract roles* is presented in fig. 12.3. As discussed in the introduction to this chapter, the key to efficient solving of this position is the observation that black king plays two abstract roles: preventing a white pawn from being promoted (by defending access to the promotion square), and defending a black pawn (the last one in a line of self-guarded black pawns). Further reasoning shows that in this situation the black king's mobility is highly restricted to a few squares only. Finally, it may be concluded that white should advance their king towards black one in order to force it to make a choice between its two roles (since at some point it will not be possible for black king to perform both its duties). The above reasoning takes advantage of the abstract roles assigned to particular pieces and groups of pieces and, at a general level, abstracts from particular pieces locations. Note, that the same analysis is applicable to positions presented in figs. 12.4(a) and 12.4(b) obtained from fig. 12.3 by moving each piece one square to the right or one square down, respectively.

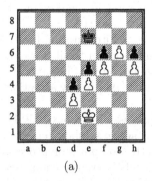
(a)

(b)

Fig. 12.4. Chess positions obtained from the position presented in figure 12.3 by moving each piece one square to the right (fig. 12.4(a)) or one square down (fig. 12.4(b)). The same *abstract roles* are assigned to pieces in all three figures (12.3, 12.4(a) and 12.4(b))

The third facet of the system proposed in [190], which strongly accounts for intuitive playing behavior is implementation of multi-level *distance-based representation of the board*. In its basic form a distance from piece X to piece

Y equals the smallest number of moves in which piece X can be moved into square occupied by Y. The above function[1] defines *relations* between pieces in a dynamic, functional way. This is one of the fundamental distinctions between discussed work and other approaches which consider the chunks of game positions as static patterns with Euclidean or Manhattan distance metric.

The distance function can be defined at different levels of accuracy. Linhares proposes the use of five such levels, starting from the very basic calculation (a heuristic glance) which omits blocking pieces, forced moves, threats, potential opponent's responses, etc. which may occur on the way, through the (level 2) estimation of a distance which takes into account blocking pieces, then estimation considering both blocks and threats by the opponent (level 3), estimation considering multiple levels of opponent responses (level 4), up to calculation based on full combinatorial search of the game tree (the highest level). Obviously the higher the level, the more time consuming the distance calculation.

Depending on the game situation (phase of the game, time remaining for making a move, strength/weakness of a position) a suitable level of distance calculation can be chosen. In many cases the final distance estimation may result from a combination of two or more distance calculation levels. Estimation of distances between vital pieces may require higher accuracy than that concerning less relevant pieces.

The multiple-level distance estimation proposed above is in line with the assumptions and experimental evidence concerning the way master chess players analyze game positions. First of all, crucial relations are examined in an exhaustive way, taking into account all possible opponents' responses, threats, forced moves, etc. The less relevant relations are only estimated as "being safe". Moreover, the dynamic, relation-based representation allows using more selective search mechanisms while analyzing a position than in the case of game representation based on static patterns. Finally, the above representation provides a plausible explanation for the relevance of empty squares in the human chunks of position representation, as observed in [59], since the distance is calculated through intermediate empty squares.

In summary, Linhares's proposal implements a cognitively plausible model of possible realization of intuitive processes in human brain and, as such, strongly deserves further attention. The approach is coherent, well-grounded, and convincingly motivated. Its main concepts are best described by the two following hypotheses underlined in [190]:

---

[1] Note that the above defined distance function is not a metric in strict, mathematical sense. Also observe, that in some cases the distance may be infinite or ill-defined, e.g. when all possible paths between X and Y are blocked by other pieces, or in the case X and Y occupy squares of different colors and X is a bishop.

Hypothesis 1: *"Human experts access chunks by the perception of abstract roles. Chunks are created when a set of abstract roles are perceived to be played by the relevant piece, groups of pieces, or squares. These abstract roles emerge from levels upon levels of subtly perceived pressures, such as pieces, empty squares, piece mobilities, attack, defence and distance relations. Chunks are composed of sets of abstract roles, and their perception leads to a strategic vision of a position."*

Hypothesis 2: *"Copycat presents a psychologically plausible architecture for chess intuition."*

In order to make the assessment fair, it is important to note that Linhares's approach is presented in principle, not as an implemented computer system. Discussions on the abstract roles played by the pieces or on the mutual competition between chunks in the short term memory, and on other facets of the system are purely theoretical and there is still a lot of work to be done on the implementation and experimental side until the proposal can be convincingly evaluated.

In particular, one of the major concerns is the changing in time role of chunks, which strongly depends on the game phase. The same chunk may have a very different contribution to the overall analysis of the position while registered in the mid-game than in the endgame situation.

Another concern is related to rather weak applicability of the proposed mechanisms in "less dynamic" games such as Go or checkers, where the notion of a multiple-level distance graph is hard to be adopted. Whereas Linhares's proposal seems to be well suited to chess (which is already a wonderful accomplishment!) the ultimate goal should be rather looking for universal approaches, applicable to the whole spectrum of games from the same genre (two-player, perfect-information, zero-sum board games).

12.5   Other Approaches

A discussion on intuition presented above in this chapter was concentrated on construction of game playing models (namely chess models), which attempted to implement intuitive behavior in a cognitively plausible way. Hence, presented methods rely on autonomous generation of game patterns and construction of hierarchical pattern libraries, usually in the form of specially designed weighted graphs.

In the literature, apart from cognitively motivated approaches, there have been published quite a few ideas of implementing the intuitive behavior in a *functional way* which means that *the effects* of their application resemble (to some extent) those of human intuitive behavior, though are accomplished without aiming at modeling human brain processes explicitly. This group includes, among others, methods focused on search-free playing or on substantial narrowing of the game tree without deep search and intensive calculations, discussed in chapter 10. Other examples of functional implementation of intuitive playing are presented in the following sections.

12.5.1   Focused Minimax Search

The idea of *focused search* or *focus networks* presented in [230] consists in limiting the number of searched possibilities in each node of the game tree. Focus networks are aimed at overcoming the two major weaknesses of the minimax search: the vulnerability to the inefficiencies of the evaluation function and the potential negative effects of the assumption that the opponent always makes the best move. Although, the former problem can in theory be alleviated by using a more sophisticated evaluation function (if available and applicable in the allotted amount of time), the latter one is an intrinsic aspect of the minimax search method. The assumption about the optimal opponent's play can be harmful (from a practical point of view) in two ways: first, it hinders taking risk by the player (since it is assumed that "tricky" though non-optimal moves would be punished by the opponent), and second, in the case the assessment of all continuations shows that the game is lost, the minimax algorithm does not promote moves which are more risky but, as such, may increase the chances of non-losing, and give hope that the opponent may not find the proper recipe.

Focus networks are developed through evolutionary process and act as filters deciding which moves in a given state look promising enough to be further explored. Only the moves which exceed a certain threshold in the initial evaluation are included in the search window. The remaining ones are discarded from the search procedure. The search continues until a predefined depth is reached, where a static evaluation function is applied and its values are propagated upwards in a standard minimax manner supported by the alpha-beta pruning algorithm.

The above idea of neuro-evolutionary focused search was applied to Othello yielding very promising results. The chromosomes coded the MLP architectures with fixed input and output layers' sizes and variable numbers of hidden units and connections. The input layer in each network was composed of 128 units organized in 64 pairs - one pair per board square, with game positions coded in the following way: $(0, 0)$ represented an empty square and $(1, 0)/(0, 1)$ black/white stone, respectively. The combination $(1, 1)$ was not allowed. The 128-unit output layer was divided into two parts - each containing 64 units (one unit per board square). The output represented the degree of network's recommendation for making a move at a given quare. Separate output units were used for network's moves and the opponent's moves. Such a distinction allowed modeling different playing styles of each side.

Following the coding used in DNA strands, the *marker-based* coding [119], described in section 6.4.3, was applied. Each chromosome consisted of 5 000 8-bit integers from the interval $[-128, 127]$ which coded the network's architecture and weights. Genetic population used in the evolutionary process consisted of 50 chromosomes among which the top-15 were allowed to mate with each other in a two-point crossover resulting in 30 new offspring in each generation. The offspring chromosomes replaced the least fitted individuals in the population. Mutation consisted in adding a random value to a mutated allele. The rate of mutation was equal to 0.4%. The best three individuals were excluded from mutation.

In order to calculate the network's fitness, it was included in the alpha-beta search algorithm and pitted against a full-width (f-w) minimax alpha-beta search. Both programs used a fixed depth $d = 2$ and employed the same evaluation function when compared. Two specific evaluation functions were used: one reflecting positional strategy (but without taking into account mobility issues) of Iago [270] - one of the first championship-level Othello programs - and one taken from Bill [182] - one of the best Othello programs in the world at that time. In order to avoid repetition of games the initial game's state was randomly selected among the 244 possible positions that can be achieved after 4 moves. Furthermore, the opponents applied ε-greedy move selection with $\varepsilon = 0.1$. The network's fitness was determined as the number of wins in a 10-game match.

The networks were evolved for 1 000 generations and then the best individual was tested against a f-w alpha-beta, both using the above described version of Iago's evaluation function. All 244 initial positions were used in the test and the opponent was not making any random moves. The focus network's search depth was fixed at 2 and the depth of the opponent varied between 1 and 5. The results showed that the efficacy of the focus network at $d = 2$ was equivalent to f-w search of $d = 4$. Moreover, the focus network's results were consistently better than the ones accomplished by a f-w minimax alpha-beta with $d = 2$.

In the second test both the focus network with alpha-beta and f-w alpha-beta adopted the same level of search depth, between 1 and 6. At each level the focus network visibly outperformed the opponent, with the highest edge achieved for $d = 2$ (the training depth). These results are especially interesting from a theoretical perspective since they show that the focus network can be trained at shallow search depth (e.g. equal to 2) and generalize well to other depths (at least within some range). The above findings stand in some opposition to Schaeffer et al.'s conclusion [288] concerning TDL-Chinook that the search depth of training games should be equal to that used in practice and suggest that shallow search approach may be valid when estimating *the relative* strength of the moves without focusing on their true numerical evaluation.

It should be emphasized that the focus network in each searched node considered, on average, only 79% of all possible moves for further examination, immediately discarding the remaining 21% of them. Since both f-w minimax and focus network applied the same static evaluation function at the leaf states and the focus network examined only a subset of all available moves, the inferior results of f-w minimax suggest that the algorithm must have been receiving some "misleading" information from the search process. Apparently this was a consequence of not always valid assumption about perfect play of the opponent, and/or the inefficiency of the evaluation function. The focus network, on the contrary, was capable of leaving out that harmful information by concentrating on the *winning* moves rather than the *safe* ones.

In the second group of experiments Bill's evaluation function was used by both players. In this case the focus network achieved a score of 51% against a f-w search. The results suggests that a very sophisticated function of Bill is hard to be improved (is close to optimum). However, it should be stressed again that in each position the focus network examined, on average, only 84% of the moves considered by Bill.

In summary, the results of Moriarty and Miikkulainen are encouraging and replicating this approach in other mind games is an interesting open issue. Focus networks behave similarly to humans in that they selectively choose the moves worth consideration and skip the ones that seem to be not promising. Experimental results show that inefficiencies of the evaluation function and the intrinsic limitations of the minimax alpha-beta search can be partly overcome in this approach. What's more, the focus networks proved to be effective also when playing with deeper search levels than the one used during training. Such flexibility is desirable in tournament game playing where time constraints need to be taken into account.

12.5.2 Linguistic Geometry and Pioneer Project

One of the facets of human game playing is the ability to abstract relevant game features from a given board position. This skill, as already discussed, allows experienced players to almost immediately estimate positional and tactical strengths and weaknesses on both sides as well as point out future possibilities and potentially promising moves. In chess, for example, these crucial features include *pawn structure*, *cooperation of figures* (e.g. two rooks on the 2nd (respectively 7th) line or multiple attack on square f2 (f7)), *mobility*, and many other.

In practically all popular mind board games vital positional and tactical features are context-sensitive. Due to the presence of other pieces on the board their appropriate classification is not a straightforward task for machine players and requires both abstraction and generalization capabilities.

A challenging test of generalization skills applicable to machines is solving game problems defined on arbitrarily large game boards. Intelligent approach to such problems requires efficient generalization from shallow search results. John McCarthy in 1998 in his comments to intelligent chess problem solving, referring to the famous *Reti problem* (fig. 12.5) in chess, stated: "Note that Reti's idea can be implemented on a 100×100 board, and humans will still solve the problem, but present programs will not ... AI will not advance to human level if AI researchers remain satisfied with brute force as a substitute for intelligence ... Would anyone seriously argue that it is impossible for a computer to solve the Reti problem by other than brute force?" [222].

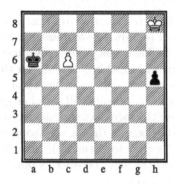

Fig. 12.5. *Reti problem.* White to begin and draw

A promising approach to this type of generalization is expressed by the Linguistic Geometry (LG) which focuses on evaluation of *trajectories of possible solutions* instead of exact exploration of the complete game tree [310]. In place of traditional search-based approach, LG proposes methods for construction of *strategies* of solving the problem. These strategies are represented as trajectories on the board (e.g. in the endgame problems) and provide a plausible combination of expert knowledge and intuition[2] (see [310] for formal description of the method). In the case of Reti problem such a trajectory-based strategy is defined along the main diagonal: Kh8-g7, Kg7-f6, ...

The ideas behind LG were initially developed by a group of scientists gathered around prof. Mikhail Botvinnik, a mathematician, but above all a former World Chess Champion, who was working on one of the first strong chess playing machines from late 1950s to the 1980s. In the 1970s Botvinnik initiated the research project *Pioneer*, which in the early stage was concentrated on investigating the methods of analyzing chess positions used by advanced chess players [37]. In particular, the focus was on development of search mechanisms that would allow substantial reduction of the game tree to be investigated.

The first version of the program was announced in 1976 and proved its applicability to certain types of positions - especially the endgames. For example, the search tree generated by Pioneer when solving the Reti problem consisted of 54 nodes, with a branching factor $b \approx 1.68$ and depth equal to 6. For comparison, the full-width game tree of depth 6 is composed of about 10^6 nodes in this case [310].

It is a pity that besides published results of Pioneer's applicability to several tactical positions little can be said about system's real playing strength since Pioneer has never publicly played a complete game of chess.

---

[2] Linguistic Geometry, in its general formulation, extends beyond game playing or game problem solving. It is generally dedicated to efficient solving of complex search problems in various domains by constructing very narrow and deep problem-related search trees.

12.5.3   Emergent Intuitive Playing

Theoretically, *the signs* of human-type intuition in machine playing may possibly emerge as a side effect of using a close to optimal evaluation function (on condition that such a function could be practically specified and implemented). Examples of "intuition" of such origin have been observed in the famous Kasparov vs. Deep Blue re-match, in which some of the machine's moves were described by grandmasters commenting on the match as "phenomenal" or "extremely human". Namely, the Deep Blue's 37th move (Be4) of game two was so brilliant that according to Kasparov and other grandmasters only the top human players could have played it, especially in the context of "natural" strong continuation 37. Qb6 (see fig. 12.6). The move 37. Be4 as well as two other "deeply human" moves, 23. Rec1 and 24. Ra3, played in this game, proved that Deep Blue was in some sense capable of exhibiting profoundly strategic human-like play, which emerged on top of the close to perfection built-in evaluation function[3].

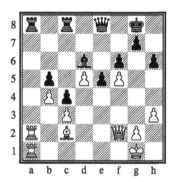

Fig. 12.6. Deep Blue vs. Kasparov. New York 1997 (game 2). After 36...axb5 the machine has played a deeply strategic, *intuitive* move 37. Be4!, despite "obvious" continuation 37. Qb6

12.5.4   Perception in Go

The considerations concerning human cognitive skills in game playing presented above in this chapter refer almost exclusively to chess, mainly due to, historically, the highest research interest attracted by this game, compared to other games.

Despite historical leadership of chess as the *queen of games*, the current interest in applying human intuitive methods in game domain will most probably be gradually switching to Go, which is potentially an even more promising application domain for this type of approaches than chess. Unlike chess,

---

[3] A very human-like way of playing in game 2 and in particular the quality of the 37th move even led Kasparov to accuse the IBM team of cheating in this game by manually adjusting parameters of Deep Blue's evaluation function.

Go is still far from being efficiently played by machines and it is very likely that, as stated by Müller in his excellent overview paper [237], "No simple yet reasonable evaluation function will ever be found for Go". The premises that support Müller's claim, already listed in chapter 4.6, include above all a requirement for complex static positional analysis, and the existence of a large spectrum of distinct, local tactical threats, which at some level of granulation need to be combined into the ultimate position evaluation. The reader interested in further exploration of this issue is directed to [237] for a more thorough discussion.

Due to the above reasons, which imply the implausibility of brute-force approach in this game, Go is very likely to gradually attract more and more research attention in the context of human-like cognitive methods. One of the first psychologically motivated research attempts in Go domain was replication of Chase and Simon's work related to chunks of information in chess discussed above. The results of experiments conducted among Go players [265] were quite different from those of chess players. Specifically, the chunks in Go, seem to be organized in the overlapping clusters, unlike in the case of chess, where they are organized in hierarchical, nested way.

In more recent works Burmeister et al. [49, 50, 51] studied the structure of Go knowledge representation of human players. The first experiment [51] was performed among Australian Go players and aimed at verifying the role of inference in the memory performance. During the experiment the task was to reconstruct the sequence of opening moves of length between 17 and 35, in two tested modes - *assisted* and *cued*. Each of these two modes was further divided into two cases: *episodic* and *inferential*.

In the assisted mode the stones were visible during the reconstruction and available for selection. In the episodic task stones were sequentially added on the board every 2 seconds and then the subject was to indicate the order in which the stones had been played. In the inferential task, only the final position was provided and the subject was to infer the order of the stones' placement.

In the cued mode the tested subjects were not assisted during the reconstruction and their task was to select the point (a crossing) on which the next move had been played. In the case of episodic task, the opening sequence was presented in a similar way as in the assisted mode, then the board was cleared, and the reconstruction phase commenced with the board being initially empty. In the inferential reconstruction case the first stone placed on the board was visible. In both tasks the subject was to infer the places for subsequent stones in the correct order of their placement.

In either mode (assisted or cued) the inferential information was the one inferred from the Go knowledge possessed by the tested subject and from the feedback information received during the reconstruction, whereas the episodic information was related to individual episodes, e.g. a single move (the placement of a stone) or the sequence of moves constituting a particular

Go opening or *joseki*[4] [51]. In all variants of the experiment subjects were not allowed to proceed to the next stone unless the current one was properly placed. They received feedback information concerning the correctness of their current attempts. The stone was placed for them after 10 unsuccessful trials of its reconstruction.

The general conclusion from the experiment was that the cued reconstruction was visibly more demanding than the assisted one. Furthermore, for all subjects the performance on the episodic tasks was higher than on the inference tasks.

The experiment was replicated a few years later in the unassisted (cued) mode on the group of Japanese Go players, leading to qualitatively the same conclusions [50]. Again the performance on the episodic tasks was higher than on the inference ones, for both experienced players and beginners.

Based on the results of both studies it was concluded that the *meaning* of the information (moves) is an important factor contributing to efficient memory organization and performance in the case of Go. This observation is in line with the findings of Chase and Simon [60, 61] with regard to comparable performance of expert and novice players in memorization and reconstruction of random chess positions, described in chapter 12.1.

12.5.5 fMRI Scanning

Interesting insights into the problem of how human cognitive abilities are implemented at neurophysiological level are presented in [10, 67]. In the first paper [10], the authors employ a functional magnetic resonance imaging (fMRI) to measure brain activity of people analyzing or observing chess boards. The subjects in the experiment were chess novice players (defined as those acquainted with the rules of the game and basic strategies that at some time had played chess regularly). Each subject was presented with three types of chess boards arranged in a specific order: 1) an empty board - at seeing which the subject was to fixate attention at the center of the board, 2) boards with randomly placed pieces (both white and black), none of them placed properly in the center of the square (the majority of the pieces partly occupied two or more adjacent squares) - the task in this case was to find the pieces (both white and black) marked by five-pointed star of low contrast color, 3) boards representing chess game positions extracted from real games, having between 25 and 30 pieces - the goal for the subject in that case was selection of the next move for white.

A similar experiment, regarding the game of Go, was conducted in [67]. Amateur Go players were repeatedly shown three types of Go boards: 1) an empty board - their goal was to fixate their attention at the center of the board image, 2) boards with randomly scattered 30 pieces (fifteen of each color), which were not properly placed, i.e. were all off the crossing points - the

---

[4] A *joseki* is a sequence of moves which results in an equal outcome for both playing sides.

goal in that case was to find the six stones (three black and three white) that were marked with a dot in a low-contrast color, and 3) regular Go positions that appeared in real games - the goal in that case was finding the next move for black.

In both chess and Go-related experiments boards were presented sequentially according to a predefined order. Each board was exposed for the period of 30 seconds. Subjects' brain activity was scanned during the whole experiment.

Based on detailed analysis of the fMRI scans collected during experiments the authors of both papers have drawn several conclusions concerning activity patterns of various brain regions in response to each of the three different kinds of stimuli respectively in chess and Go. The most interesting, however, seems to be the comparative analysis aimed at specifying the similarities and differences in brain activities for both games. In particular, two of the main research questions were "Are the same [brain] areas involved in playing Go and chess?" and "Are there different degrees of lateralization in Go and chess?" [67]. Considering deep differences between the two games (Go is a spatially organized territory game with no pieces differentiation other than the color, with complicated hierarchical position analysis, whereas chess is a game better suited for straightforward analysis, with well defined goals, and with diversity of pieces of different strength and striking range) quite relevant differences between the respective fMRI scans were expected.

Surprisingly, the scans were not as varied as one might have predicted. On a general level the same regions were activated when solving tasks in either of the two games. The only noticeable difference was the involvement of one more region (the left dorsal lateral prefrontal area) - usually responsible for language functions - when focusing on Go positions. This area was generally inactive in the case of chess positions. One of the hypothetical explanations is that subjects were verbalizing Go terms and specific stone combinations when analyzing Go boards.

The other general distinction between the results was the observation that in the case of chess the left hemisphere showed moderately more activation than the right one, and in the case of Go the opposite situation was registered, the right hemisphere was moderately more active than the left one [67]. These differences to some extent confirm that Go is a more pattern-oriented game than chess, which in turn relies more on analytical calculations.

In summary, the idea of conduction comparative studies aimed at tracing the similarities and differences in human brain's activation for both games is definitely an interesting research avenue. It should be noted, however, that the results presented in both papers are only preliminary and several other experiments need to be preformed in order to state firm conclusions. Furthermore, when interpreting such results one need to be very cautious about establishing a proper relation between the reasons and the effects. The observations concerning higher activation of particular brain areas may be interpreted based on the roles currently attributed to these regions, but it is also possible that these observations actually discover new properties of those areas.

12.6 Summary

Herbert Simon, one of the giants working on intuition, argued that intuition is nothing mysterious or extraordinary and simply relates to a subconscious pattern recognition process able to immediately provide appropriate pattern(s) among those stored in the memory, based on our knowledge and experience [299]. According to Simon, this does not mean that intuition is an irrational process - he considered it to be a rational but neither conscious nor analytical one. Simon was optimistic about potential implementation of human cognitive skills in computers and claimed that AI had already reached the level at which intuition could be modeled [300].

Understanding and furthermore implementing the mechanism of intuition in artificial players is one of the main challenges for CI in games. Several issues described in this book, e.g. search-free playing or feature abstraction and generalization may partly contribute to the implementation of intuition, but efficient and universal approach to this wonderful human ability is yet to be specified. Certainly, unless the mechanisms driving human intuition are known in detail, the research concerning human-type intuition in game playing will be to some extent speculative, not fully experimentally verifiable. This does not mean, however, that looking for plausible implementations is not worth the effort. On the contrary!

In 1993 Atkinson [11] wrote that "The master's superior play is due to 'sense of position', an intuitive form of knowledge gained from experiencing a great variety of chess situations. Intuitive knowledge is perhaps the most important component of expertise in any human discipline, yet intuition has been one of our least understood phenomena. Intuition is an automatic, unconscious part of every act of perception. ... Intuition is nonverbal, yet trainable. Intuition is not part of present-day machine chess."

Following the above statement, it could be argued that unless programs (machines) capable of making intuitive moves (in the above described sense) in chess and other mind board games are created, one should be very cautious about announcing the end of the human era in these games. The above manifesto does not mean that, in the author's opinion, it is expected that traditional brute-force approaches would become inferior to selective intuition-based methods in popular board games. Honestly speaking, the attempts to mimic human cognitive abilities in mind games made to date did not keep pace with the hardware-oriented brute-force search approaches. However, one can reinterpret these observations and say that it is only a matter of sufficient game's complexity for the selective intuitive methods to come into play or even take the lead - *vide*, for example, the game of Go or special variants of chess.

Certainly, the problem of whether machines can possess intuitive skills or be creative extends far beyond the game domain. Many complex real-life problems require partly intuitive approaches. Actually, in everyday life people in most cases do not base their decisions on exhaustive logical analysis of all options. When solving complex problems some fraction of intuitively non-promising paths is immediately discarded. As Duch stated in [85]: "Intuition is manifested in categorization based on evaluation of similarity, when decision borders are too complex to be reduced to logical rules. It is also manifested in heuristic reasoning based on partial observations, where network activity selects only those paths that may lead to solution, excluding all bad moves."

Potential success of intuition-focused research in games will be a significant step towards general understanding of this still mysterious human cognitive ability.

13

Creativity and Knowledge Discovery

Creativity, besides intuition, is another hallmark of human-level intelligence. There exist several definitions of this competence, but all of them generally agree that creativity is inseparably tied with novelty and selectivity. According to Sternberg and Lubart [309] creativity can be defined as "the ability to produce work that is both novel (...) and appropriate (...)." Another possible definition discussed in [36] is formulated in terms of "novel combination of old ideas," where the novelty of a creative solution is reflected by its non-obviousness and interestingness. Yet another dictionary definition is "to bring into being or form out of nothing."

The question of whether computers can be creative accompanied their development from the very beginning. Lady Ada of Lovelace, a friend of Charles Babbage (the inventor of Analytical Engine implementing early ideas of universal computing machine), wrote in her memoirs in 1842, referring to Babbage's proposals, that: "The Analytical Engine has no pretensions to *originate* anything. It can do *whatever we know how to order it* to perform" [74, 326]. In 1950 Turing, in his seminal paper on machine intelligence (in which he among other things proposed the Imagination Game known today as the Turing Test) [326], recalled Lady Lovelace's objections and responded to her disbelief that machines could be creative: "It is quite possible that the machines in question had in a sense got this property. [...] The Analytical Engine was a universal digital computer, so that, if its storage capacity and speed were adequate, it could by suitable programming be made to mimic the machine in question." Turing concluded the paper by summarizing his point of view on machine intelligence and creativity: "We may hope that machines will eventually compete with men in all purely intellectual fields," and pointed out the game of chess as the field on which the above efforts should be initially concentrated.

The question of whether computers can be creative and what does it really mean to be creative in the case of machines is still not addressed in a commonly accepted way. Machine creativity can be discussed in various perspectives including first of all philosophical and psychological viewpoints.

J. Mańdziuk: Knowledge-Free and Learning-Based Methods, SCI 276, pp. 205–214.
springerlink.com   © Springer-Verlag Berlin Heidelberg 2010

Such a thorough discussion, although definitely useful and needed, is beyond the scope and capacity of this book.

Leaving out theoretical considerations, it could be generally agreed, however, that despite having or not the ability of being creative in a human sense, the machines are potentially capable to computationally *simulate* the process of being creative, i.e. to accomplish the *functional effects* of creativity, albeit, not necessarily by following human cognitive models. In this perspective, creativity in games can manifest itself as discovery of new game features, game openings, or playing styles. In particular, one of the very tempting research tasks is the problem of autonomous generation of game-specific features that may constitute efficient evaluation function.

The question of *how can a system generate useful features automatically?* dates back to 1960s. It was raised by Samuel in his pioneering work on checkers [278, 279], but at that time left without satisfying answer. The set of basic features used by Samuel's checkers program was specified by the author and the aim of the learning phase was to develop a suitable set of coefficients in the linear combination of these features leading to strong evaluation function [277]. This approach was improved in the revised versions of the system, but still the basic set of primitive components had to be defined by human expert.

Selected examples of autonomous knowledge discovery in games are presented in the remainder of this chapter. In the first group of them, features discovered by the playing system originated in a spontaneous way, as a "side effect" of the training procedure, rather than as a result of a deliberately planned process (e.g. in the case of TD-Gammon or Blondie). The other group of examples consists of attempts representing "planned discovery," i.e. systems, such as Zenith, Morph, or Moriarty and Miikkulainen's coevolutionary approach, which were designed already with the focus on being creative, i.e. capable of discovering new game-related knowledge, which may be effectively used in autonomously defined evaluation function. Some of the examples refer to systems which have already been presented in previous chapters. In such cases their description is very brief and restricted to the aspect of knowledge discovery.

13.1 *Spontaneous* Knowledge Discovery

13.1.1 TD-Gammon

The most famous example of creativity in games is probably TD-Gammon described in chapter 6.1, which, according to former backgammon world champion Bill Robertie, came up with genuinely novel strategies that no one had used before. TD-Gammon's play caused revision in human positional judgement in the game leading, for instance, to invention of new opening moves. The openings proposed by TD-Gammon were subsequently proven to be successful in exhaustive, statistical analysis as well as in tournament play.

Another interesting observation concerning TD-Gammon was a development of spatial patterns in its weight space responsible for representation of particular game concepts, which were not explicitly presented in the course of training [317]. Detailed analysis of weights of the best two-hidden-layer TD networks revealed that the first hidden layer units developed into race-oriented feature detectors whereas the second hidden layer ones became attack-oriented feature detectors. This specialization was manifested by specific weight patterns (e.g. spatially oriented linear gradation of negative/positive weights) explainable by experienced backgammon players.

13.1.2 DDBP Neural Network Solver

Similar observations regarding *ad-hoc* feature discovery and feature representation in neural network's weights were reported in [209, 233, 234, 235] concerning the Double Dummy Bridge Problem.

Several MLP networks described in chapter 8.4 were trained in a supervised manner and tested based on the data from the GIB Library [133]. Weight patterns discussed below have been found in all types of tested networks and for both *notrump* and *suit* contracts (the networks were trained separately for these two types of deals). For the sake of simplicity, the description presented below addresses only the case of 52-input deal representation (described in chapter 8.4) and *notrump* contracts. The observations were repeated over the ensemble of several such architectures.

Recall that in the 52-input-based deal representation the input layer was composed of 52 neurons and each of them was assigned to a particular card from a deal. Neuron's input value denoted a hand containing the respective card ($N : 1.0$, $S : 0.8$, $W : -1.0$, $E : -0.8$). A single output neuron yielded the predicted number of tricks to be taken (the output range - $[0.1, 0.9]$ was divided into 14 intervals of equal length). Besides deal assignment, no additional information e.g. the rules of the game or the strength of particular cards was provided to the network. Apart from achieving promising numerical results the other main goal of that research was exploration of networks' knowledge representation and search for patterns in their weight spaces that possibly represented particular bridge features (e.g. the relative strength of cards). Examining the weights of trained networks revealed several interesting observations.

Figure 13.1 presents visual representation of connection weights in a typical $52 - 25 - 1$ network trained on *notrump* contracts. Each connection weight is represented as a black (for negative value) or white (for positive value) circle with radius depending on the weight's value. The leftmost, separate column represents weights of connections from hidden neurons (numbered to the left of the column) to the output neuron. The main area of circles represents weights of connections from all 52 input neurons (assigned to cards from a deck, as depicted below the area) to 25 hidden neurons (numbered in the leftmost column). For example the connection weight from the 15th input

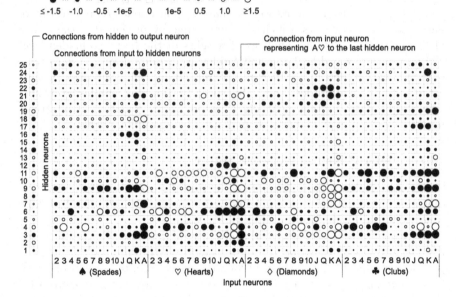

Fig. 13.1. Visualization of connection weights of the $52-25-1$ network trained on *notrump contracts*. Each circle represents one weight. Connection weights between hidden neurons and the output neuron are placed in the leftmost, separate column, and those between input and hidden neurons in the remaining rectangular area. The radius of each circle represents the absolute value of the respective weight. Black circles denote negative weights, and white positive ones

neuron (representing 3♡) to the 6th hidden neuron is positive and greater or equal 1.5 (the largest possible empty circle), whereas the connection weight from the 6th hidden neuron to the output one has a small negative value (close to zero), represented by a small black circle in the leftmost column.

The first observation from the figure is the existence of hidden neurons with seemingly "random" values of input connections and very small absolute values of output connections, e.g. neurons number 6, 10 or 24. Such neurons seem to be useless, but they are apparently important - when they were pruned out from the network the performance significantly worsened.

Another pattern found in the figure, is the concentration of relevant connections (with biggest absolute values of weights) on honors (i.e. *Aces, Kings, Queens, Jacks,* and *Tens*). Additionally, for each investigated network, it was possible to point out exactly 4 connections with absolute values much greater than the others. All these "favored" connections were linked to input neurons assigned to *Aces*. In the network presented in fig. 13.1, these connections were the following: for A♠ to hidden neuron number 9 (with value 24.19), for A♡ to neuron number 3 (-26.89), for A♢ to neuron number 11 (26.71), and for A♣ to neuron number 4 (26.69). For comparison, the greatest absolute value among all the remaining connections was equal to 6.47. This phenomenon is easy to explain (for humans): *Aces* are the most important cards in the game of bridge, especially in *notrump* contracts.

The third weight pattern also emphasizes the importance of honors in the game of bridge. For each suit, one hidden neuron specialized in honors of that suit can be selected: the hidden neurons number 16 (for ♠), 12 (for ♡), 22 (for ♢), and 17 (for ♣). All these 4 neurons follow the same weight pattern: only connections from input neurons representing honors of the respective suit have significant values. All the other weights are much smaller. It is very interesting that for these neurons not *Aces*, but *Queens* and *Kings* are the most important. Even *Jacks* are more important (for these neurons) than *Aces*. For human bridge players it is obvious that the presence of all figures makes the suit much more powerful and simplifies taking tricks (especially in *notrump* contracts where there is no possibility to ruff) without a need of a *finesse* (i.e. playing based on a hypothetical assumption about which of the opponents possesses the missing figure).

Another pattern is revealed when input connections to hidden neurons with numbers 2, 8, 18, and 19 are compared. As in the previous patterns, each of these neurons specializes in one suit, respectively ♡, ♢, ♠, and ♣. This specialization is manifested by favoring all cards of the respective suit and *Aces* from the other suits - all these cards have values of connection weights of the same sign, and connections from all the other cards have the opposite sign. Furthermore, the importance of cards in the relevant suit is graded according to the rank of the card: the *Ace* has the biggest absolute value of connection, the *King* relatively smaller, etc. It is interesting that for these four hidden neurons the importance of *Aces* of other suits is smaller than the importance of the *two* of the specific suit (however still noticeable).

Certainly, such observations might be treated as occasional, if they didn't exist in all tested architectures, trained independently. Moreover, qualitatively similar conclusions were drawn for other deal representations, and for *suit* contracts, as well.

All the above observations are in line with human knowledge about the game of bridge. Estimation of strengths of individual cards as well as the entire suits is the basic information considered in the process of hand's evaluation. Even though these game features are trivial to understand for human players, they are not necessarily easy to discover by a neural network. In a simple training process, the networks were also able to autonomously discover the notion of *the finesses* - a subtle manoeuvre quite often deciding about the final number of tricks taken by a playing pair.

Another interesting phenomenon observed exclusively for *suit* (*spades*) contracts was visible relevance of the trump suit cards. After training with simplified $52 - 1$ architectures (with 52 inputs, one output, and no hidden units) the three most important cards in a deal were $A♠, K♠$, and $Q♠$. $A♡, A♢$, and $A♣$ were less important than $Q♠$ (the *Queen* of trumps) and $K♡, K♢$, and $K♣$ were less important (i.e. had smaller absolute values of outgoing connections) than $8♠$. This observation is, once again, in line with human conviction of particular usefulness of trump suit cards compared to cards from other suits in the case of *suit* contracts.

13.1.3 Blondie

A well-known example of spontaneous feature discovery in games is Blondie24, also known as Anaconda [64, 65, 109] described in chapter 6.2, which received its name due to the "snake-like" way of playing. In many games won by the program its opponent was blocked and therefore forced to make a weak move. However, neither in the input data nor in the evolutionary process of Blondie24/Anaconda's development the concept of mobility was ever explicitly considered. Hence the importance of mobility must have been "invented" by the system in the course of evolution.

The above accomplishment can be partly attributed to the hand-designed topology of the input-to-hidden connections (c.f. chapter 6.2, figure 6.1), however, the mutual spatial relations between subsections of the board covered by individual neurons in the first hidden layer were never explicitly defined in the self-play data, *ergo* were discovered through evolution.

In 2005, a Blondie-like approach was applied to Othello by Chong et al. [70]. The authors essentially followed Chellapilla and Fogel's methodology, which was only formally adapted to the specificity of the game (see section 6.4.1 for a detailed description). A similar effect of spontaneous discovery of crucial game features as in the game of checkers was observed. The general conclusion was that coevolutionary method represented by Blondie was able to evolve players having the sense of both material advancement and positional nuances. Yet, picking up a mobility strategy appeared to be a much more demanding learning task, but given enough evolution time of 1 000 generations some individuals managed to discover the importance of mobility and were able to properly combine all three aspects (material, positional, and mobility) into a unified evaluation system.

13.2 *Deliberate* Knowledge Discovery

In the systems discussed in the following sections the issue of discovering new game features has been stated as one of the explicit goals already at the stage of system's design. In other words, knowledge discovery mechanisms are implemented in these systems intentionally, aiming at autonomous development of the game features. Some of these methods attempt to extract the basic game features from a board position with the help of deliberately designed decomposition processes or based on feature induction mechanisms. Others choose a bottom-up approach and strive to build complex sophisticated game features out of predefined set of basic, *a priori* designed primitive features. The process of construction is again unguided although its underlying mechanisms are in most cases planned by the system's designer.

Four examples of such deliberate knowledge discovery attitude are presented below. Naturally, they do not exhaust the list of possible approaches. However, in the author's opinion, they compose a meaningful set of representative ideas in this matter. In this context the selection key was diversity of discovery mechanisms driving these systems.

13.2.1 Morph

Interesting observations concerning knowledge discovery in chess were reported in the Robert Levinson's Morph experiment [141, 185, 188]. Morph was designed according to Levinson's *credo* that computational creativity should be implemented as unique and favorable combination of past experiences, which is in line with "novel combination of old ideas" viewpoint. The system implemented pattern based learning with the weights of patterns being modified through the TD(λ) method combined with *simulated annealing* (pattern's individual learning rate was decreasing along with the increase of the frequency of pattern's updates).

Patterns learned by Morph were consistent with human chess knowledge. In particular Morph was able to play openings at a reasonable level, even though no information about the significance of development or controlling the center of the board in the opening phase had been explicitly provided to the system. Some of the Morph's moves could be considered creative, especially when taking into account its shallow 1-ply search. Examples of such moves include [186]: "winning a piece through removal of a defender" or "offering a pawn or piece to achieve a strong attack and then regaining the material" or even "playing to checkmate the opponent's king in a losing position rather than passively defending."

One of the weaknesses of Morph was poor scalability with respect to the number of patterns, due to the lack of efficient selection mechanisms. Although the strength of Morph was far inferior to GNU Chess (being its teacher), the system was nevertheless able to defeat human novices while searching 1-ply only.

13.2.2 Zenith

A general approach to the problem of feature discovery in games was presented by Fawcett and Utgoff in Zenith system [101, 102, 103, 104]. Being initially equipped with the specification of the problem to be solved, Zenith was capable of autonomous generation of various new problem-related features. Starting from an initial set of predefined *goal-oriented* features (possibly including only one element - the goal of a game) the system applied four types of feature transformations: *decomposition, goal regression, abstraction,* and *specialization*.

An example of *decomposition* transformation given by the authors for the game of Othello was the following: the player wins when no moves can be made by either side and the player has more discs than the opponent. The latter condition can be *decomposed* into two sub-features (subgoals), i.e. calculating the number of discs belonging to the player, and calculation of the number of them belonging to the opponent. Measuring and providing these two parameters separately to the learning system is more powerful than operating on their difference or any other combined measure.

The *goal regression* transformed the existing features by regressing them through the domain operators. For example, since the goal of the player is to maximize the number of his own discs, it seems desirable to verify the number of discs that will belong to the player after a move (or a sequence of moves) or alternatively to measure the number of potentially acquired/lost discs.

Abstraction was applied to features in order to lower the cost of their verification during position evaluation. This transformation relied on skipping specific details and in effect simplifying the original feature and making it more general, albeit without sacrificing too much of its accuracy.

The fourth type of transformation was *specialization*, accomplished by looking for feature invariants, i.e. the special cases (e.g. domain elements or their configurations) in which conditions of that feature were always satisfied. Instead of testing a set of expensive conditions of the original feature the system created a new, simplified feature and verified the existence of these special cases, which was both faster and less expensive than testing the conditions of the original feature.

The feature generation process was combined with the *concept learner*, responsible for selection of the optimal set of features - labeled as *active* - to be used in the evaluation function. The remaining features were labeled *inactive* (at that moment), but might be included in the evaluation function at a later time.

Application of the system to Othello described in [101, 104] included nine types of transformations (four *decompositions*, three *specializations*, and one type of *abstraction* and *goal regression*, respectively). The system was initialized with a single feature: *the win of black*. Applying iteratively and selectively the above-mentioned nine transformations allowed "discovery" of some well-known *positional* game features used by advanced human players, e.g. the notions of X-Squares and C-Squares, stable and semi-stable discs, as well as several mobility features. More importantly, through the multiple transformation process Zenith also discovered an entirely new, previously unpublished game feature, called by the authors *future mobility*, related to the moves that would become available in the next state. On the other hand, it is fair to mention that some relevant game-features were inaccessible by Zenith, i.e. could't be reached in proposed iterative transformation process.

It is worth emphasizing that unlike "typical" approaches to the feature generation problem which seek efficient feature combinations, Zenith was oriented on accomplishing particular goals in the considered domain and therefore the game features were designed in the functional rather than structural way.

The main advantage of Zenith was the universality of its feature generation mechanisms, which were applicable to a wide class of games. It is a pity that the work on Zenith was not pursued further and the system was not verified in other board games[1].

In summary, Zenith represented one of the most significant attempts of autonomous feature discovery and further investigation of this direction is desirable and potentially beneficial. The ultimate goal to be accomplished on this path is development of an autonomous system able to yield features as efficient as those used by top human experts.

13.2.3   Discovering Playing Strategies in Othello

Another example of creative learning in games is Moriarty and Miikkulainen's approach to Othello [231] described in detail in section 6.4.3. The authors implemented a specially designed coevolutionary approach with the aim of discovering appropriate playing strategies against particular types of opponents.

In the initial phase, when trained against a random player, the system discovered the importance of positional strategy, similarly to novice human Othello players. In the subsequent training against a more demanding alpha-beta player the system's strategy evolved and gradually included advanced elements of mobility.

The mobility features discovered by the evolved networks are indispensable for high-profile play and generally unattainable by novice human players.

13.2.4   Analysis of Archive Game Records

Recently Miwa, Yokoyama and Chikayama have proposed [229] a new method for generating game features to be used in the evaluation function. The features are constructed as conjunctions of predefined primitive/base features chosen either by hand or in an automatic way.

In the application to Othello presented in [229] there were 192 base features defined, each of them referring to a particular board square and representing one of the three possible situations: the square was occupied by black, or was occupied by white, or was empty. A conjunction of such primitive features composed a game feature (also called a pattern). A pattern was considered present in a given board position if all base features defining it were present in that position.

The training data set used to generate game features was composed of 200 000 game positions and the test set contained over 900 000 positions. All training examples were extracted from game records archived on a game server. Each position contained 60 discs and was labeled with black player's result of the game. The label was defined based on completing the game by searching to its end.

---

[1] It was, however, applied to another domain, namely telecommunications network management [101, 104].

The complex features built upon the basic ones were chosen according to the *frequency of occurrence* in the training games and with respect to *conditional mutual information* they provided. The frequent features were defined as the ones that appeared in the training set at least some predefined number of times. In order to alleviate computational load of the method, features (patterns) that were similar to each other, based on some similarity measure, were combined into one feature. Further selection of patterns relied on maximization of the conditional mutual information of the newly selected pattern for every pattern already chosen, using the CMIM (Conditional Mutual Information Maximization) method [108, 229].

In the experimental evaluation, the proposed method selected several thousand game features and achieved 77.2% accuracy on the test set in the win/loss position classification. This result was successfully confronted with three classifiers that used base features directly, i.e. Naive Bayesian Classifier - 73.9% accuracy, an MLP trained with RPROP method - 75.2%, and Linear Discriminant Classifier - 73.4%.

13.3 Summary

All the above-mentioned systems are capable to discover new game features, previously unknown to the system, induced from the training data. In some of these approaches, feature discovery resulted from a deliberately designed learning process, with an explicit focus on creating relevant game features. In the others, game features were "unintentionally" developed in the course of learning, for instance, as distributed patterns in the weight space of neural network-based evaluation function.

Some of the presented systems are widely-known milestone achievements in intelligent game playing, but even though, they are still quite far from being truly creative. With a single exception of TD-Gammon's invention of new opening ideas, none of the game features or game properties discovered by these systems could actually be considered revealing. In this perspective, it can be concluded that the accomplishments in development of creative game playing systems made to date, are yet unsatisfactory.

The ultimate goal that can be put forward in this context is *autonomous* discovery of *all* relevant components of the evaluation function in sufficiently complicated mind game in a way allowing their *separation* and *explanation*. In particular, such a requirement extends beyond the neural or neuro-evolutionary solutions proposed hitherto, which resulted in efficient *numerical approximation* of the board state, but generally lacked *explicit feature-based formulation* of the evaluation function.

Naturally, the problem of creativity in mind games represents only one dimension of a much broader and highly challenging task of implementation of creativity mechanisms in intelligent systems, in general. The reader interested in this subject may be willing to consider [74] or [84] for further reading.

Multi-game Playing

One of the ultimate goals of AI/CI research in games is development of a universal playing agent, able to play virtually any game as long as it knows its rules. Realization of this task requires designing general-purpose learning and reasoning methods that abstract from particular games. A potential variety of possible games to be played (theoretically there are infinite number of them) with different equipment (boards, card decks, various types of moving pieces, different goals, etc.) makes the task - in its general form - extremely demanding. Therefore, on a general level the goal is yet far from being accomplished, however, some steps in this direction have already been successfully taken.

Multi-game playing became a hot topic in the beginning of 1990s when some renowned methods and systems originated. In order to make the problem tractable, these first attempts were often restricted to a certain class of games, usually two-player, perfect-information, zero-sum, deterministic ones.

The majority of approaches developed in this area rely on symbolic, logical game representations, rooted in the mainstream of traditional AI. Relatively fewer examples can be found within the CI-related methods. Selected attempts, which originated in both AI or CI areas are briefly summarized in this chapter.

14.1 SAL

One of the first widely-known CI-based universal learning agents was Michael Gherrity's SAL (Search And Learning) system [131] capable of learning any two-player, perfect-information, deterministic board game. SAL consisted of a kernel that applied TD-learning combined with neural network's backprop algorithm in order to learn the evaluation function for a given game. The kernel was game-independent and remained unchanged for different games. The rules of making valid moves for any particular game were represented by a game-specific module. The system used only 2-ply search supported by the *consistency search* method [21, 22] appropriately modified by Gherrity.

J. Mańdziuk: Knowledge-Free and Learning-Based Methods, SCI 276, pp. 215–229.
springerlink.com   © Springer-Verlag Berlin Heidelberg 2010

SAL generated two evaluation functions, one for each playing side which allowed learning nonsymmetric games or imposing asymmetry in symmetric games, if necessary. Both evaluation functions were represented as one-hidden-layer MLPs. The current game situation was characterized by some generally-defined binary features (applicable to a wide spectrum of games), which included *positional* features - based on the board position (e.g. a type of a piece on each square, the number of pieces of each type, etc.), *nonpositional* features - based on the last move made (the type of a piece moved, the type of a piece captured), and *rule-based* features (e.g. pieces potentially lost, squares potentially lost, possible win of a game, etc.). Since all features were binary, the number of them, especially in the case of varied pieces' types, might be quite significant.

The size of the input layer depended on the number of game features generated for the game. The size of the hidden layer was arbitrarily chosen as being equal to 10% of the input size. Each move made during the game represented one training example. The target value for the output represented the evaluation of the next board position and was calculated with the TD(λ) method. Neural networks were trained with off-line backpropagation algorithm, i.e. after the game was completed.

SAL's playing rate was strongly hindered by slow learning. For example, it took the program 20 000 games to learn to play tic-tac-toe. In the case of connect-4, SAL required 100 000 games to achieve approximately 80% winning rate over the training program (the details about the training program's strength are not available), using 221 game features.

In a more serious attempt to learn how to play chess, SAL played 4200 games against GNU Chess, drawing 8 of them and losing the remaining ones. During the experiment GNU Chess was set to make a move within one second, which was approximately equivalent to 1500 − 1600 ELO rating. SAL was using 1031 input features and searched to 4-ply depth on average.

The analysis of games played against GNU Chess showed that SAL's learning process had been extremely slow, but on the other hand a stable progress had been observed from the initial random play towards a more organized way of playing. An indication of this performance increase was the average length of games played by SAL (before being mated), which steadily increased from about 15 moves in the initial phase, to about 30 on average at the end of the experiment.

The question whether SAL would be capable of learning how to play chess (or another complicated game) at decent level within a reasonable amount of time remains open, but the answer is likely to be negative.

14.2 Hoyle

Another interesting approach to game-independent learning was represented by Hoyle system [93, 94, 97, 98, 99] devised by Susan Epstein. Hoyle was able to learn any two-player, deterministic, perfect-information game defined on finite board, given only the rules of the game. It used shallow search, 2-ply at most. The system employed the *lesson and practice* training scheme, discussed in section 9.1.2, which interleaved playing games against the expert (lessons) with knowledge-based self-playing periods (practicing) [98].

The underlying idea of Hoyle was to use a set of game independent Advisors, each specializing in a narrow, specific aspect of game-playing (e.g. material advantage, or finding the winning moves or sequences of moves, etc.). Each of the Advisors might recommend some moves and all of them could comment on these proposals from their specialized viewpoints. The Advisors were arranged in three tiers. Those in the first tier relied on shallow search and were focused on providing a perfect opinion about single moves, postulating their selection or avoidance. For example, the Victory Advisor mandated immediately winning moves whereas Sadder Advisor was focused on avoiding immediately losing moves [96, 97]. The tier-2 Advisors advocated certain plans of play, i.e. sequences of moves leading to achieving particular goals. Finally, tier-3 Advisors voted for (or against) particular moves based on their heuristic assessment within the Advisor's specific focus. The Advisors in tier-1 and tier-2 made their decision sequentially, i.e. if any of them was able to select the next move it did so, and such a decision could not be canceled by subsequent Advisors. The Advisors in the last tier made their decision in parallel using a simple weighted arithmetic voting scheme.

Hoyle's decision making process was highly sensitive to the choice of weights assigned to Advisors in the third tier. These weights were learned with the use of PWL (Probabilistic Weight Learning) algorithm. PWL was run after each game completed with the expert teacher and adjusted the weight of each Advisor according to the extent to which the Advisor manifested the game expertise and game knowledge, reflected by the expert's moves, in the just completed contest [97]. All game states in which it was the expert's turn to move were analyzed one-by-one in the context of supporting or opposing Advisors' comments related to the recorded expert's move. Based on these partial assessments the cumulative adjustments of the Advisors' weights were made. Consequently, each weight learned by PWL represented the probability that the Advisor's opinion was correct.

The diversity of Advisors played a crucial role in learning a new game. Each of the Advisors learned patterns from played games, picked according to its individual priorities. One Advisor might be focused on opening move patterns, while another one, for instance, on those related to strong, winning moves, etc. Besides game-specific knowledge that could be acquired by the Advisors from analysis of the played games, Hoyle was *a priori* equipped with some general knowledge about the domain of two-person, deterministic games.

Hoyle was presumably the first AI system with confirmed *ability to learn more than one game*. Its efficacy was demonstrated in 18 two-player board games, including tic-tac-toe, lose tic-tac-toe and nine men's morris [97, 98].

The potential of Epstein's approach in more complicated games has not been experimentally proven.

In the spirit of this book it is worth emphasizing that Hoyle's learning and decision making processes exhibited some similarities to that of people: Hoyle tolerated incomplete and inaccurate information and was able to consider several conflicting rationales simultaneously, during the move selection process. It integrated pattern-based learning with high-level reasoning. Hoyle considered the obviously strong (winning) and obviously poor (losing) moves first (in tier-1), before the remaining ones. A decision in tier-3 was generated as a result of negotiation process between various, partly conflicting concepts. Finally, Hoyle's decisions were explainable through a natural language interpretation of Advisors' comments.

Taking into account the above-mentioned principles underlying Hoyle's design, the substantial research effort related to system's development, and favorable experimental results, there is no doubt that Hoyle is one of the milestone achievements in multi-game playing.

14.3 METAGAMER

METAGAMER proposed by Barney Pell [248, 249, 250] represented another approach to multi-game playing. The system was applicable to a class of symmetric chess-like (SCL) games. According to [252] "A symmetric chess-like game is a two-player game of perfect information, in which the two players move pieces along specified directions, across rectangular boards. Different pieces have different powers of movement, capture, and promotion, and interact with other pieces based on ownership and piece type. Goals involve eliminating certain types of pieces *(eradicate goals)*, driving a player out of moves *(stalemate goals)*, or getting certain pieces to occupy specific squares *(arrival goals)*. Most importantly, the games are *symmetric* between the two players, in that all the rules can be presented from the perspective of one player only, and the differences in goals and movements are solely determined by the direction from which the different players view the board." Typical examples of SCL games include chess, checkers, draughts, Chinese-chess, or shogi.

METAGAMER was equipped with a minimax-based search engine with alpha-beta pruning and iterative deepening heuristic. The heart of METAGAMER was game-independent, universally-defined evaluation function composed of a set of predefined simple features (goals). Based on the rules of a particular game provided as the input, the system constructed efficient game representation and suitable evaluation function to be used by a generic search engine. Human predefined knowledge was restricted to general description of the SCL game framework. Playing any particular representative of this class required game-specific optimization, which did not involve human intervention.

Similarly to Hoyle, each feature in the evaluation function was defined in the form of an advisor deciding whether this particular game aspect was advantageous for the player or for his opponent. Unlike in Hoyle, however, the METAGAMER advisors could formulate only positive opinions, i.e. their statements were, by definition, only favorable. A negative opinion could be formed by stating a favorable opinion with respect to the opponent's situation. Each advisor was defined by a specific heuristic rule, which returned appropriate value to be added to the overall estimation of position's strength. Four groups of advisors were implemented concerning *mobility*, *threats and capturing*, *goals and step functions*, and *material*, respectively. Each of the first three categories of advisors had 4 representatives and the last one included 11 advisors [251, 252].

METAGAMER was used by Pell [251] for material analysis, that consisted in estimation of the relative strength of checkers (man vs. king) and chess pieces. This analysis was performed based exclusively on the rules of the game (respectively checkers and chess) with the help of the previously-mentioned 11 *material* advisors. Quite surprisingly, the estimated values of some chess pieces were close to the figures commonly used by humans[1]. The values of a knight and a bishop were comparable (as expected) and equal to 46.9 and 51.7, respectively. The value of a rook was equal to 75.5. In this context a pawn and a queen, worth 12.6 and 103, respectively, were both underestimated[2].

In checkers, the relative strength of a man (ordinary checker) vs. a king was within the expected range. The respective estimations were equal to 13.9 and 23.2, which yielded the king-to-man ratio of 1.67, whereas the common heuristic suggests that this ratio should belong to the interval $[1.5, 2.0]$.

In preliminary experiments METAGAMER was pitted against 1992's version of Chinook and against GNU Chess (version A.D. 1992). Pell concluded [251, 252] that METAGAMER was about even to Chinook playing at the easiest level, when given a one man handicap. Thanks to advisors focused on mobility, METAGAMER was able to "rediscover" the value of a strategy of not moving its back men until late in the game (moving the back man makes the respective promotion field more easily accessible for the opponent).

In the case of chess, METAGAMER appeared to be comparable to GNU Chess, when given a handicap of a knight and when GNU Chess was playing at level 1, i.e. searched 1 ply with possible extensions in non-quiescent positions. METAGAMER searched 1 ply with occasional extensions to the depth of 2.

---

[1] Certainly, the values of particular chess pieces are highly position-dependent, and therefore proposed numbers reflect only statistical estimations of their average strengths.

[2] In order to be consistent with the estimated values assigned to rooks, knights and bishops, queens and pawns should be worth approximately 150 and $15 - 20$, respectively.

Overall, the METAGAMER experiment represented an interesting approach to building universal game-playing system, albeit restricted to a certain class of games. One of the major unresolved problems was construction of the set of weights assigned to advisors or, as in the case of Hoyle, the autonomous mechanism of negotiating/updating the weights. This issue was left as an open problem for further research, but to the author's knowledge the work on METAGAMER has not been continued or at least its results not published, since 1996 Pell's paper. In preliminary tests all weights were set to 1, but certainly, similarly to Hoyle, a mechanism of weights adaptation to particular game would be strongly desirable when striving to build a universal, efficient game player.

It would also be interesting to observe the METAGAMER's performance in the case of playing some *ad-hoc* defined SCL games, which would ultimately prove the quality of this concept. Some preliminary results of such contests against random players and against other versions of METAGAMER were reported in [250, 252], but there is still room for more experimental evaluations.

In the perspective of knowledge-free approaches, both Hoyle and METAGAMER were not pure representatives of this paradigm. On the one hand their evaluation functions were defined in a general, largely game-independent way. On the other hand the definitions of advisors were certainly biased by the human knowledge. For example, METAGAMER's *mobility* advisors implemented well-known static and dynamic features applicable to classical mind board games. Since the construction of advisors partly reflected human domain knowledge about SCL games it is quite unlikely that they would be easily applicable to another or broader class of games.

14.4 Morph II

Morph II, developed by Robert Levinson [187], was yet another renown example of game-independent learning system and problem solver. Morph II, also called Universal Agent, was a direct continuation of Morph - an adaptive pattern-oriented chess learning system mentioned in section 13.2.1. Morph II combined CI techniques with AI symbolic learning. It used neural network-like weight propagation learning and genetic pattern evolution. Game rules and game patterns were represented as *conceptual graphs*, which allowed fast and efficient incremental update of the pattern database. The system was capable of autonomous abstraction of new features and patterns, and development of its own learning modules. Like its predecessor, Morph II relied on shallow search equal to only 2 ply, on average. While Morph has successfully learned to play chess at a novice level, the objective of Morph II was reaching the strength of a master level player, which is equivalent to 2200 ELO ranking points or above.

Morph II was designed as a domain independent system. Domain patterns or features were not supplied to it, but derived from the game (domain) rules using a small set of generally-defined mathematical transformations. The weights assigned to these features (patterns) were not provided to the system, as well, but autonomously determined by Morph II. The learning procedure was parameterless. All learning coefficients and parameters were dynamically adjusted based on the system's experience.

The objectives of Morph II project were very ambitious and apart from achieving a master level of play in chess, included also "achieving significant (expert-level or higher) performance in a domain other than chess" as well as exhibiting "reasonable strength (determined experimentally and in competition), after training, across a large class of state-space search problems - including new ones to be developed" [187].

In this perspective, Morph II research project directly addressed the issue of designing a domain-independent, multi-task learning system, but its performance in domains other than chess was not demonstrated.

14.5   General Game Playing Competition

The idea of General Game Playing (GGP) proposed at Stanford University [129, 130] follows and further extends the multi-game playing research directions. According to [129] "General game players are systems able to accept declarative descriptions of arbitrary games at runtime and able to use such descriptions to play those games effectively (without human intervention)." The class of games considered in GGP is extended, compared to previous approaches, to *finite, synchronous, multi-player* games. "These games take place in an environment with finitely many states, with one distinguished initial state and one or more terminal states. In addition, each game has a fixed, finite number of players; each player has finitely many possible actions in any game state, and each terminal state has an associated goal value for each player. The dynamic model for general games is synchronous update: all players move on all steps (although some moves could be "no-ops"), and the environment updates only in response to the moves taken by the players" [129].

Finite, synchronous games can, technically, be represented as state machines. Their rules are defined in first-order logic, with some extensions. Game descriptions are provided with the use of Game Description Language (GDL) [193] in the form of logical sentences that must be true in every state of the game. GDL is a formal language for defining discrete games of complete information, potentially extendable to the case of games with incomplete information. Formal description of GDL and an example of coding the rules of the tic-tac-toe game can be found in [193] and [130].

Since games' descriptions are formulated in the language of logic, the most natural way to implement the GGP agent is with the use of logical reasoning. A computer player can, for example, repeatedly interpret game description at each step of the game playing process. Alternatively game description can be mapped onto another representation, which is then interpretively used during the game. Another possibility is to apply some predefined programming procedures which, based on a game description, can autonomously devise a specialized program capable of playing the game. Other approaches are also possible.

Designing a competitive GGP agent is a truly challenging task requiring innovative solutions in several AI/CI areas, including efficient knowledge representation and transfer between learning tasks, automated reasoning in multi-agent environment, domain-free heuristic search, effective opponent modeling, and example-based learning and generalization.

Cluneplayer

The most successful in the first two editions of annual GGP Competition, were systems relying on automatic generation of game features based on a predefined set of universal, generic features [71, 292]. These generated features were combined into an evaluation function reflecting the specificity of the currently played game.

The first GGP champion was *Cluneplayer* [71], the winner of the 2005 contest. Cluneplayer through analysis of game's description abstracted the game to its three core aspects: *expected payoff, control* (relative mobility) and *expected game termination*. These game aspects were modeled as weighted linear combinations of the (automatically selected) game-state features. The feature selection process was performed based on the three predefined types of interpretations applied to the game rules. In effect, a set of candidate features was defined and their ultimate selection was guided by the, appropriately defined, *stability* of each candidate feature.

Depending on the complexity of the game (and the time allotted for game analysis) some of the aspects appeared to be more relevant than others in the final form of the evaluation function. In more complex games (Othello, chess, or six-player Chinese checkers) Cluneplayer was unable to discover control or termination features, reducing the evaluation function to the combination of payoff-related terms and weights. In the case of a much simpler game - race-track corridor (introduced in the First Annual GGP Competition), described in [71], all three game aspects were developed and quantified by the system during the pre-game analysis period.

Fluxplayer

The *Fluxplayer* [292], the 2006 winner, determines the degree to which certain formulas in a game description support the predicates defining *goal* and *terminal* states. These estimates are calculated with the use of fuzzy logic by assigning values of 0 and 1 (or more generally $1 - p$ and p, for fixed $p \in [0, 1]$) to atoms, reflecting their truthfulness in a goal state, and using T-norm and associated S-norm to calculate the degree of truth of complex formulas. The method is further enhanced by automatic detection of structures in a game description and assigning some non-binary values, other than p and $1 - p$, as their truth evaluation. The structure detection part of the method follows the ideas proposed in another GGP-related paper by Kuhlmann et al. [175] with some vital modifications. The general playing policy of Fluxplayer is to avoid terminal states unless the goal is accomplished. Consequently, the value of predicate *terminal* has positive contribution to the evaluation function if the value of *goal* is high, and negative, otherwise.

CadiaPlayer

Quite a different approach to GGP is adopted by the winner of 2007 and 2008 competitions - *CadiaPlayer* [107]. The GGP agent proposed by Finnsson and Björnsson uses a Monte Carlo (MC) simulation-based approach enhanced by the UCT algorithm (Upper Confidence bounds applied to Trees) described in section 3.3.1.

The UCT method maintains the game tree in memory and keeps track of the average return of each state-action combination $Q(s, a)$ that has been played. The action a^* to be explored in the currently considered state s is chosen according to the following formula [107]:

$$a^* = argmax_{a \in A(s)} \left\{ Q(s, a) + C \sqrt{\frac{\ln N(s)}{N(s, a)}} \right\}, \qquad (14.1)$$

where $Q(s, a)$ is the action value function, $A(s)$ denotes the set of all actions a available in state s, $N(s)$ is the number of times state s was visited, and $N(s, a)$ denotes the number of times action a was explored in state s.

The UCT enhancement to the MC simulations allows establishing an appropriate balance between exploration and exploitation of the game space. Paths in the game tree which are not promising are explored less frequently than the good lines of play, consequently the latter ones grow much more aggressively in the maintained game tree.

Besides applying state-of-the-art game tree search algorithms, CadiaPlayer also uses opponent modeling techniques, in which a separate model is assigned to each player in multi-player games.

On a general note, the MC-UCT method has recently become very popular in game community. The method appeared to be especially well suited to Go and several currently leading developments in this game are related to particular MC-UCT realizations (see section 3.3.1 for more details).

The main advantage of applying simulation techniques lies in their ability to implicitly capture game properties in real time. Furthermore, the MC-UCT simulations are easy to parallelize promising taking full advantage of multi-core massively-parallel machines in the future.

On the downside, simulation-based methods, despite being guided by previous goodness of considered moves are, at their core, implementing an extensive search procedure and as such are - anyway - limited by the computing resources and available time, especially in the case of complicated games or the ones defined on larger boards (e.g. 50×50 or 100×100 Go).

NEAT-Based Player

Due to limited amount of time allotted before the start of the tournament for the playing system's preparation (including acquisition of game rules), which is insufficient for executing a full learning or self-learning procedure, the use of CI methods in GGP is severely hindered, especially in the case of more complicated games. Despite potential difficulties, it could be argued that approaching GGP with Computational Intelligence methods seems to be worth consideration. One of the main reasons for pursuing CI research in this area is possibility of coming up with new ideas related to generalization and transfer of knowledge acquired during previous phases of the tournament to the new, unseen games, which would be beneficial in the GGP formula as a multi-game tournament.

One of the promising applications of CI to GGP, based on coevolution is proposed in [264]. The authors use the so-called NEuroevolution of Augmenting Topologies (NEAT) method [307], which allows simultaneous evolution of a neural architecture and its weights. Each genome is composed of a list of *connection genes* specifying each connection as a 5-tuple $< from\_node, to\_node, weight, enable\_bit, innovation\_number >$. The *enable_bit* specifies whether the connection is enabled (active) and the *innovation_number* is a specific connection label, which allows finding the corresponding connection during the crossover operation without the need for extensive topological search.

Evolution in the NEAT method starts from small and simple networks and enlarges them as needed in the coevolutionary process with speciation. Speciation allows the topologically new structures to be developed to their optimized form before being forced to compete against individuals from other niches in the population.

Each neural network in the population represents a heuristic evaluation function. Initially each such network is represented as a simple Perceptron with 40 inputs and one output, without hidden units. The input to the network represents a random projection of the set of input features (associated with the current game state) onto the 40 input neurons. In each generation an ensemble of neural networks compete with one another for promotion to the next generation. Due to the time constraints the lookahead search is restricted to 1 ply only.

In the coevolutionary fitness assessment process each individual is evaluated against some combination of the opponents drawn from the population, according to the *covering competitive algorithm* (CCA) [271]. CCA maintains two populations (say A and B), and in each of them a ranked set of previously dominating strategies, called teachers (denoted by T_A and T_B, respectively). Teachers $t_A \in T_A$ and $t_B \in T_B$ are ordered according to the time (generation number) they were developed. A population (either A or B) is said to be currently dominated if no individual capable of beating all teachers from the opponent population has been developed in this population yet. At each generation, each individual $a \in A$ plays games against teachers t_B in the order respective to the teachers' ranks in T_B. If a loses any game, no more games are played. If a is a winner against all $t_B \in T_B$, it is introduced to T_A and population B is denoted the currently dominated one. Finally, at the end of the evolutionary process, the highest ranked member of the teachers' set (for both populations) is chosen as the ultimate solution.

In addition to the CCA evaluation, each individual also plays some number of games against randomly chosen opponents from the current generation. Fitness is ultimately calculated as a weighted sum of the scores obtained during the CCA and in the above complementary evaluation against peers.

The method was tested on five two-player games from the GGP corpus: connect-4, a simplified version of Chinese checkers, two-board tic-tac-toe and two other less popular games. The efficacy of proposed method differed significantly between the games. The detailed results of the learning progress (measured by the teachers' set size), the scalability of the solution (understood as its applicability in the case of deeper lookahead searches), and the disengagement problem (a situation in which one of the populations outperforms the other one to a degree which hinders possible search for further improvement) are thoroughly discussed in [264].

Generally speaking, in most of the cases the evolved player outperformed a random opponent, which suggests that the coevolutionary approach was capable of extracting some relevant information related to played games. On the other hand, as stated by the authors, there are still several directions for future improvement of the method. Some of them are concerned with design of the coevolutionary process in general, other address the problem of a more effective inclusion of game domain knowledge (in the current system's implementation the first-order logic game description is not transformed to the set of game features, and doing so seems to be one of possible enhancements).

Although several issues, e.g. limited scalability or disengagement problem, need further investigation, the overall assessment of the Reisinger et al.'s approach is definitely positive, especially when taking into account that this work represents one of the very first attempts of applying CI-based methods to GGP domain.

Summary

The General Game Playing Competition is undoubtedly a challenging task for both symbolic, logical AI methods and evolutionary or neural-based CI approaches. Playing effectively various *a priori* unknown games, even when restricted to a certain class of games, requires using universal game representation schemes, self-adaptation of the goal function, incremental knowledge acquisition and its efficient sharing and generalization through subsequent tasks.

One of the highly demanding issues in GGP is the possibility for game playing agents to take advantage of playing multiple times the same game against various opponents or different games against the same opponent. Theoretically it might be possible to discover and exploit some characteristic features of particular games or some weaknesses in the opponent's playing style. In practice, however, the above idea is hard to follow since the players are not aware beforehand of the name of the game (only the rules of the game are presented) and the names of the opponents. Without such "labeling" a player must discover similarities between played games and similarities in the opponents' playing styles that indicate the presence of a particular game or opponent, based exclusively on its internal mechanisms. Such a task is very hard to accomplish even for humans, which are generally much smarter than machines in this type of "blind" identification.

Another research avenue concerns further development of learning algorithms that maintain an appropriate balance between exploration and exploitation. One of the well-known possibilities, implemented in the current champion program (CadiaPlayer), is to use the MC-UCT simulations. Another promising option is proposed by Sharma et al. [297] based on a combination of TD(0) and Ant Colony Optimization (ACO) [83]. The method performs random simulations and uses the TD-learning to assess visited states according to the probability of winning from those states. The choice of paths to be explored during simulations depends on the past assessments combined with the so-called *pheromone* and *desirability* values of the states defined by the ACO method.

The GGP Competition is mainly devoted to board perfect-information games. One of the future prospects may be an attempt to integrate various types of games - board and card games, perfect and incomplete information ones, as well as deterministic and chance games, as proposed in [260]. At the moment, however, the GGP in its current formulation seems to be demanding enough to remain a challenge for the near future.

14.6   Transfer of Knowledge

One of the key issues in building intelligent multi-task learning systems is the problem of efficient knowledge representation allowing its transfer and sharing between several learning tasks. Such a knowledge transfer can be realized within multi-task or incremental learning frameworks.

Multi-task learning (e.g. [58]) utilizes simultaneous learning of a few tasks and sharing representation issues, experience and knowledge among them in order to make the overall learning process faster and more effective. *Incremental learning*, on the other hand, is usually implemented as a lifelong, sequential learning process, i.e. tasks are learned one after another, although representation of problems as well as knowledge acquired in previous learning are widely shared between tasks and involved in subsequent learning, consequently making the learning progressively easier (e.g. [214, 324, 325, 339]).

In the domain of games, the mainstream research was, historically, focused on designing agents which excelled in particular games, they were devoted to, but were completely useless in playing other, even somehow similar games. The best-known examples are Deep Blue and Chinook, the man-machine world champions in chess and checkers, respectively. Both systems were *a priori* equipped with formidable amount of domain knowledge, including large opening and endgame databases and sophisticated hand-crafted evaluation functions designed by human masters in the respective games. Achieving the world-class level of play was in these cases accomplished at the expense of using game-specific approaches, which were nontransferable to other games in any automated or semi-automated way.

In the 1990s the ideas related to multi-game playing came into light, which brought about universal learning/playing systems designed to pick up any game within a certain class of games, such as Zenith, SAL, Morph II, Hoyle, METAGAMER, and several other systems not mentioned in this book. These systems were able to learn how to play a new game using predefined but, in principle, game-independent mechanisms. The common ground for all of them was, however, treating each game as an entirely new learning task, not linked to previous learning experience. Consequently, even though a set of learning tools used by each of these systems was common for all learned games, learning tasks were totally separated.

The ideas related to sharing of knowledge (experience) between games became a hot research topic only in the last few years. The resurgence of interest in this research area was mainly caused by the increasing popularity of the GGP Competition. In the case of games belonging to the same genre (e.g. two-player, perfect-information board games) several representation issues and meta-level game rules/decriptors are either common or similar. Hence, theoretically speaking, learning a new game does not have to start from the very beginning. The learning system may exploit already possessed knowledge - albeit its appropriate tuning will usually be required.

Using the experience gained in learning one or more games in order to alleviate the effort of learning another, similar one resembles a learning attitude typical for humans. Previous experience in playing card games, for example, is usually one of the fundamental factors in efficient learning of a new card game. The whole "toolbox" related to the suit and card names, their hierarchy, the idea of trick completion and taking, etc. is already possessed and there is no need to thoroughly repeat this knowledge acquisition process - some limited adjustments are sufficient in most of the cases. Certainly, a description of formal rules and goals of any particular game is a separate issue and may differ significantly from previous experiences.

In the game literature there are very few examples of the methods concerned with task-to-task knowledge sharing and transfer. One of possible meta-heuristics in this context is to rely on analysis of the rules of a newly considered game - in an automated way - and find commonalities or close similarities to previously learned games. Parts of the evaluation functions (features and weights) defined for prior games, associated with these shared features, can then be used as a first-step heuristic approximation of the evaluation function for the current (new) game.

An example of CI-based approach of this type is presented in [16], where the authors developed a TD(λ) learner able to automatically discover game tree template structures of height two (2-ply deep) acting as game features, and acquire their respective assessments. The features are described in a game-independent way and as such are potentially transferrable between games from a considered genre of GDL-based games. The set of acquired features together with accompanying assessments is used as a starting point when learning a new game. Although it is assumed in the paper that a new learning task uses exactly the same set of features as was used in the previous task, in a more general approach one may think of gradually extending the set of features and initializing them with prior values, if available, or default ones, otherwise.

The proposed method was tested in the tasks of learning connect-3, captureGo, and Othello, based on the features and their associated values acquired from tic-tac-toe. In all three cases the knowledge-based TD-player outperformed the baseline TD-learner which was learning from scratch and did not take advantage of the tic-tac-toe features. Knowledge-based learning proved to be both faster and more effective.

Another possible approach to knowledge transfer between the learning tasks concerns the case of two versions of the same game, differing by the size of a game only. The basic idea is to learn relevant game features in a simplified game environment and then apply this knowledge to speed-up and enhance the learning process in the desired game's version (size). Examples of such methods are presented in chapter 8.3. Certainly transfer of knowledge within the same game seems to be a significantly simpler task than the case of different (though "similar") games, but still the experience gained from these approaches may be beneficial in designing the "truly" multi-task learning systems.

14.7 Summary

Artificial Intelligence established itself as a serious research discipline about 60 − 70 years ago and since the very beginning posed the question of how to define and measure intelligence in machines. Although several functional definitions of machine intelligence have been proposed, none of them is widely agreeable, actually. Some people opt for the Turing test as a measure of intelligence, others argue that in order to pass the Turing test a machine does not have to be intelligent - it only needs to be "smart enough" to mislead human judges.

Most of the descriptions of machine intelligence agree, however, that one of its hallmarks is the ability to learn how to perform previously unknown tasks in a new, competitive environment. The question which is still open is whether a truly intelligent machine is the one which can improve its behavior in one task until the theoretical limits of optimal performance are reached or, on the contrary, an intelligent machine is the one which can learn efficiently how to solve many different, possibly unrelated tasks, but does not have to surpass human competitors in all of them - more or less in the same way as any human individual is not universally good at performing all possible tasks and may be very efficient in solving some of them, but at the same time poorer (below average) in some other. The thesis that intelligent behavior requires high-level specialization in performing certain (intelligent) tasks is confronted in AI/CI with the claim that a truly intelligent agent is (similarly to people) multi-purpose and able to consider new tasks within its area of expertise.

The latter point of view is shared by the author of this book, who believes that intelligence in machine playing means, among other properties, *being universal*, i.e. capable to learn various new games and take advantage of experience and knowledge gained from prior learning tasks while learning a new one.

All systems presented in this chapter are examples of multi-task approaches to machine intelligence. All of them are general learning systems, potentially capable of picking up any game within a certain class of games. All have demonstrated the ability to learn how to efficiently play simple games or more complicated ones, but at a novice level only.

Definitely, multi-game playing research deserves further exploration and the problem of how to design universal, game-independent learning systems, capable of playing many various games at a decent level (competitive to intermediate human players) is one of the grand challenges in the area of intelligent game playing, and consequently in machine intelligence in general.

15

Summary and Perspectives

Artificial Intelligence, since its very early stages has been highly focused on games. Usually described by a compact set of rules, with clear goals and well-defined terminal states, games provide cheap, replicable environments, perfectly suited for testing new learning methods and search algorithms.

Among various types of games, the mind games (especially two-player, perfect-information, zero-sum board ones) seem to be particularly tempting for AI. For many years, the ultimate goal of approaching and then surpassing humans in these widely popular and intellectually demanding tasks used to be another source of their popularity. The paradigm examples of games considered by AI were chess, checkers, Othello, and Go, which attracted game-related research groups and individuals all over the world. The human-computer competitions in the first three games ended with decisive victories of machines in 1997 (chess: Deep Blue), 1994 (checkers: Chinook) and 1997 (Othello: Logistello), respectively. Among the most popular mind board games only Go remains as the last stronghold of human supremacy, but recent advances of the Monte Carlo UCT algorithm suggest that the human predominance over machines in Go is likely to gradually diminish. Poker and bridge, widely popular card games are, besides Go, two more examples of games which are still not dominated by computer players. The recent AI/CI advances in both of them have, however, also been very telling.

The common feature of most state-of-the-art game playing systems is the use of several game-specific enhancements, which are not straightforwardly transferrable to other games. These include hand-crafted evaluation functions with weight coefficients tuned manually by top human players, extended opening books and endgame databases, specific game-related improvements in the tree search methods, or in some cases also high-end specialized hardware.

The cumulative effect of applying the above sophisticated techniques led to building very powerful game playing systems stronger than the best human players. In particular, the victory of Deep Blue over Kasparov had a huge impact on mind game research and marked the end of era of narrowly focused

J. Mańdziuk: Knowledge-Free and Learning-Based Methods, SCI 276, pp. 231–234.
springerlink.com   © Springer-Verlag Berlin Heidelberg 2010

research aimed at outplaying human world champions "at any cost." This event confirmed the thesis that there is limited sense in pursuing this type of research any further, and most probably end up building even more powerful dedicated machine capable of even faster search and equipped with even better hand-tuned evaluation function. Certainly, the research focused on extending the margin between machines and humans in chess is still continued (yielding, for instance, a wonderfully playing system Rybka), but the main goal on this path has been accomplished and there is virtually no chance that humans will ever become a threat to machines again. A similar situation is the case in checkers, where after Chinook's success over Tinsley, its developers have gone further ahead and recently solved the game. In face of the above circumstances the mind game research community had no choice but switch its priorities and define new goals and perspectives for the field.

Generally speaking, approximately in the beginning of the 1990s, two new trends in mind games have originated. One stream of research concentrated on application of Computational Intelligence to development of strong playing programs devoted to particular games, but either with the use of knowledge-free approaches or in limited-domain-knowledge regime. Among CI subdisciplines the most intensive research was focused on application of evolutionary methods, neural networks, reinforcement learning, or combinations of the above. These three areas composed a strong basis for development of universally applicable learning algorithms, often taking advantage of the effect of synergy. The most notable examples include TD-Gammon system implementing TD-learning paradigm combined with neural representation of the evaluation function, which accomplished a world-class level in the game of backgammon, or Blondie24 neuro-evolutionary approach to checkers which attained a level of playing competency comparable to master human players. Interesting results were also obtained by Moriarty and Miikkulainen with their neural network-based, search free approach to Othello without the use of domain knowledge.

Simultaneously, another significant trend in artificial mind game playing - focused on development of multi-game learning mechanisms - gained momentum. One of the first renowned examples of such systems were Zenith, SAL, Morph II, Hoyle, and METAGAMER. All of them relied on shallow search (just 1 or 2 ply) and were capable of picking up any game within an *a priori* defined class of games. All these systems had the ability to learn the games autonomously, with only limited use of domain knowledge, and to play them at a novice level or slightly above, except for very simple games, like tic-tac-toe, in which an expert competence could be demonstrated. This line of research has been recently revitalized by the annual General Game Playing Competition, which extends and formalizes the idea of multi-task learning within a certain class of games.

It seems reasonable to expect that putting together these two research streams, i.e. application of general purpose CI/AI methods and tools in the quest for developing highly competitive systems, each devoted to playing a particular game, and using universal learning techniques in a multi-game learning framework, will lead, in longer perspective, to development of a universal game player equipped with several skills, hitherto reserved for humans. The list of these skills includes intuitive playing, abstraction and generalization of possessed knowledge, creativity and knowledge discovery, and multi-task and context-sensitive learning. Further development of the above skills should ultimately lead to building a domain-independent, autonomously learning machine player capable of high performance in many games across various game genres, with no requirement for initial domain knowledge other than the game rules.

In order to achieve this goal several specific problems have to be addressed on the way. Some of these challenging issues are introduced and argued for in parts III and IV of the book. The challenges and open questions presented in part III are more "technically oriented" and refer to problems related to: efficient representation of game positions and the associated evaluation function, methods of move ranking based on shallow search or with no search at all, modeling the opponent (especially in the context of uncertainty provided by games with imperfect information). Special place on this list is devoted to examining the efficacy of various TD-training schemes and their comparison with evolutionary methods. These two learning paradigms, usually combined with trainable neural representation of the evaluation function, are the main platforms for building autonomous CI learning systems. Moreover, several specific problems within TD-learning are also worth investigation, e.g. playing against external opponents vs. self-playing. Both these learning paradigms have already been successfully applied to various types of board games and proved to be effective in autonomous, experience-based learning, but at the same time some examples of failures have also been reported in the literature for both training schemes.

The most demanding research areas in the domain of mind games, according to the author's subjective opinion, are those presented in the last part of the book. The most formidable challenge is implementation of intuition in game-playing systems or, more precisely, implementation of mechanisms that would efficiently mimic human-type intuitive behavior. Such achievement would straightforwardly lead to efficacious search-free pre-selection of moves, almost instantaneous estimation of position strength, as well as the ability to play strong positional moves relying on shallow search only. All three above-mentioned skills are typical for advanced human players, but still evidently unattainable for machines.

The second of the three most demanding issues listed in part IV is implementation of mechanisms of autonomous knowledge discovery that would lead to creation of new game features and new playing strategies. In particular a very challenging task is autonomous generation of board features that compose efficient (close to optimal), descriptive game representation allowing adequate evaluation of board positions. At the moment, development of a world-class playing program requires that the set of features be defined by human experts. Even though there exist a few notable examples of learning how to play certain games without human expertise, there is still a lot of work ahead, especially in the case of more demanding games.

The third of highly challenging issues concerns game independent learning, in particular incremental learning methods allowing sequential or simultaneous learning of several games. Sequential incremental game learning may rely on appropriate tuning of already possessed knowledge and generation of new features only when necessary, i.e. when they are "sufficiently different" from already discovered features. Simultaneous learning requires the representational and computational issues to be shared online among various learning tasks (games) and each learning task benefits from this synergy. Recently, a new resurgence in this research area has been observed due to growing popularity of the GGP Competition, but in spite of several interesting propositions presented by the participants there is still a lot of room for development and verification of new, innovative ideas, before genuinely universal, strong multi-game playing systems will appear.

Certainly, the list of challenging problems described in this book is by no means complete. It is also clear that achieving even all of the challenging goals listed here, would not be equivalent to construction of an omnipotent, unbeatable artificial player, capable of playing "in God's way." As humans make mistakes and are not infallible, also the human-like playing systems, which would emerge at the end of this research path, might possibly suffer from "human" weaknesses. The above conclusion, on the other hand, does not mean that application of CI methods to mind game playing is not interesting or advantageous. On the contrary, it could be argued that mind game playing research will be more and more tied with Computational Intelligence methods and its future will be closely related to the development of psychologically motivated learning methods attempting to follow higher-level human competencies.

I am aware that each of the above three grand goals is extremely demanding and for many readers achieving anyone of them may seem to be illusive or at least very improbable. But hadn't the hitherto advancement of AI/CI in mind games been really amazing and hardly expected as well? It is my strong belief that one of the central AI/CI challenges of designing autonomous, efficient multi-task learner which, to some extent, is able to functionally mimic the human way of learning, reasoning and decision making is achievable in the context of mind games. I am also convinced that the experience gained on the way will be transferrable to other domains and helpful in achieving similar effects in other research areas.

References

1. Abramson, B.: Expected-outcome: a general model of static evaluation. IEEE Transactions on Pattern Analysis and Machine Intelligence 12(2), 182–193 (1990)
2. Akl, S.G., Newborn, M.M.: The principle continuation and the killer heuristic. In: Proceedings of the ACM Annual Conference, pp. 82–118 (1977)
3. Aleksander, I.: Neural networks - evolutionary checkers. Nature 402(6764), 857 (1999)
4. Alemanni, J.B.: Give-away checkers (1993),
   http://perso.wanadoo.fr/alemanni/give_away.html
5. Allis, V.: Searching for Solutions in Games and Artificial Intelligence. PhD thesis, University of Limburg, Maastricht, The Netherlands (1994)
6. American Checkers Federation. Checkers rules,
   http://www.acfcheckers.com/
7. Anantharaman, T., Campbell, M., Hsu, F.-h.: Singular extensions: Adding selectivity to brute-force searching. Artificial Intelligence 43, 99–109 (1990)
8. Araki, N., Yoshida, K., Tsuruoka, Y., Ysujii, J.: Move prediction in Go with the Maximum Entropy Method. In: Proceedings of the 2007 IEEE Symposium on Computational Intelligence and Games (CIG 2007), Honolulu, Hawaii, pp. 189–195. IEEE Press, Los Alamitos (2007)
9. Arbiser, A.: Towards the unification of intuitive and formal game concepts with applications to computer chess. In: Proceedings of the Digital Games Research Conference 2005 (DIGRA 2005), Vancouver, B.C., Canada (2005)
10. Atherton, M., Zhuang, J., Bart, W.M., Hu, X., He, S.: A functional MRI study of high-level cognition. I. The game of chess. Cognitive Brain Research 16, 26–31 (2003)
11. Atkinson, G.: Chess and Machine Intuition. Ablex Publishing, Norwood (1993)
12. Auer, P., Cesa-Bianchi, N., Fischer, P.: Finite-time analysis of the multiarmed bandit problem. Machine Learning 47(2/3), 235–256 (2002)
13. Baba, N., Jain, L.C., Handa, H. (eds.): Advanced Intelligent Paradigms in Computer Games. Studies in Computational Intelligence, vol. 71. Springer, Heidelberg (2007)

14. Bahçeci, E., Miikkulainen, R.: Transfer of evolved patter-based heuristics in games. In: Proceedings of the 2008 IEEE Symposium on Computational Intelligence and Games (CIG 2008), Perth, Australia, pp. 220–227. IEEE Press, Los Alamitos (2008)

15. Baker, R.J.S., Cowling, P.I.: Bayesian opponent modeling in a simple poker environment. In: Proceedings of the 2007 IEEE Symposium on Computational Intelligence and Games (CIG 2007), Honolulu, Hawaii, pp. 125–131. IEEE Press, Los Alamitos (2007)

16. Banerjee, B., Stone, P.: Ganeral Game Playing using knowledge transfer. In: Proceedings of the 20th International Joint Conference on Artificial Intelligence (IJCAI 2007), Hyderabad, India, pp. 672–677 (2007)

17. Barone, L., While, L.: An adaptive learning model for simplified poker using evolutionary algorithms. In: Proceedings of the Congress of Evolutionary Computation (GECCO 1999), pp. 153–160 (1999)

18. Baxter, J., Tridgell, A., Weaver, L.: Experiments in parameter learning using temporal differences. ICCA Journal 21(2), 84–99 (1998)

19. Baxter, J., Tridgell, A., Weaver, L.: Knightcap: A chess program that learns by combining td(λ) with game-tree search. In: Proceedings of the Fifteenth International Conference on Machine Learning (ICML 1998), Madison Wisconsin, July 1998, pp. 28–36 (1998)

20. Baxter, J., Tridgell, A., Weaver, L.: Learning to play chess using temporal differences. Machine Learning 40(3), 243–263 (2000)

21. Beal, D.F.: An analysis of minimax. In: Clarke, M.R.B. (ed.) Advances in Computer Chess 2, pp. 103–109. Edinburgh University Press, Edinburgh (1980)

22. Beal, D.F.: Recent progress in understanding minimax search. In: Proceedings of the ACM Annual Conference, pp. 165–169. Association for Computing Machinery, New York (1983)

23. Beal, D.F.: A generalised quiescence search algorithm. Artificial Intelligence 43, 85–98 (1990)

24. Beal, D.F., Smith, M.C.: Learning piece values using temporal differences. ICCA Journal 20(3), 147–151 (1997)

25. Berliner, H.: The B* tree search algorithm: A best-first proof procedure. Artificial Intelligence 12(1), 23–40 (1979)

26. Berliner, H., Ebeling, C.: Pattern knowledge and search: the SUPREM architecture. Artificial Intelligence 38(2), 161–198 (1989)

27. Beyer, H.-G.: The Theory of Evolution Strategies. Natural Computing Series. Springer, Berlin (2001)

28. Beyer, H.-G., Schwefel, H.-P.: Evolution Strategies: A comprehensive introduction. Natural Computing 1(1), 3–52 (2002)

29. Billings, D.: Thoughts on RoShamBo. ICGA Journal 23(1), 3–8 (2000)

30. Billings, D.: International RoShamBo programming competition (2001), http://www.cs.ualberta.ca/~darse/rsbpc.html

31. Billings, D., Davidson, A., Schaeffer, J., Szafron, D.: The challenge of poker. Artificial Intelligence 134, 201–240 (2002)

32. Billings, D., Paap, D., Schaeffer, J., Szafron, D.: Using selective sampling simulations in poker. In: Proceedings of the British Spring Symposium on Search Techniques for Problem Solving under Uncertainty and Incomplete Information, pp. 13–18 (1999)

33. Bishop, C.: Neural Networks for Pattern Recognition. Oxford University Press, Oxford (1995)
34. Blair, A.: Learning position evaluation for Go with Internal Symetry Networks. In: Proceedings of the 2008 IEEE Symposium on Computational Intelligence and Games (CIG 2008), Perth, Australia, pp. 199–204. IEEE Press, Los Alamitos (2008)
35. Bledsoe, W.W., Browning, I.: Pattern recognition and reading ny machine. In: Proceedings of the Eastern Joint Computer Conference, pp. 225–232 (1959)
36. Boden, M.: Creativity and computers. In: Dartnall, T. (ed.) Artificial Intelligence and Creativity: An Interdisciplinary Approach. Studies in Cognitive Systems, vol. 17, pp. 3–26. Springer, Heidelberg (1994)
37. Botvinnik, M.M.: Computers in Chess: Solving Inexact Search Problems. Springer, New York (1984)
38. Bouzy, B.: Associating domain-dependent knowledge and Monte Carlo approaches within a Go program. Information Sciences 175(4), 247–257 (2005)
39. Bouzy, B.: Associating shallow and selective global tree search with Monte Carlo for 9x9 Go. In: van den Herik, H.J., Björnsson, Y., Netanyahu, N.S. (eds.) CG 2004. LNCS, vol. 3846, pp. 67–80. Springer, Heidelberg (2006)
40. Bouzy, B., Cazenave, T.: Computer Go: an AI oriented survey. Artificial Intelligence 132(1), 39–103 (2001)
41. Bouzy, B., Chaslot, G.: Bayesian generation and integration of K-nearest-neighbor patterns for 19×19 Go. In: Proceedings of the 2005 IEEE Symposium on Computational Intelligence in Games (CIG 2005), Colchester, UK, pp. 176–181. IEEE Press, Los Alamitos (2005)
42. Bouzy, B., Chaslot, G.: Monte-Carlo Go Reinforcement Learning experiments. In: Proceedings of the 2006 IEEE Symposium on Computational Intelligence and Games (CIG 2006), Reno, NV, pp. 187–194. IEEE Press, Los Alamitos (2006)
43. Bouzy, B., Helmstetter, B.: Monte Carlo Go developments. In: van den Herik, H.J., Iida, H., Heinz, E.A. (eds.) Proceedings of the Advances in Computer Games 10 (ACG 10), pp. 159–174 (2004)
44. Breuker, D., Uiterwijk, J., Herik, H.: Replacement schemes for transposition tables. ICCA Journal 17(4), 183–193 (1994)
45. British Othello Federation. Othello rules, http://www.britishothello.org.uk/rules.html
46. Brudno, A.L.: Bounds and valuations for abridging the search of estimates. Problemy Kibernetiki 10, 141–150 (1963) (in Russian)
47. Brügmann, B.: Monte Carlo Go. Technical report, Max-Planck-Institute of Physics, Germany (1993)
48. Bump, D.: GNU Go (1999), http://www.gnu.org/software/gnugo/gnugo.html
49. Burmeister, J.: Studies in human and computer Go: Assessing the game of Go as a research domain for cognitive science. PhD thesis, School of Computer Science and Electrical Engineering and School of Psychology. The University of Queensland, Australia (2000)
50. Burmeister, J., Saito, Y., Yoshikawa, A., Wiles, J.: Memory performance of master Go players. In: van den Herik, H.J., Iida, H. (eds.) Games in AI research, pp. 271–286. Universiteit Maastricht, Maastricht (2000)

51. Burmeister, J., Wiles, J.: The use of inferential information in remembering Go positions. In: Matsubara, H. (ed.) Proceedings of the Third Game Programming Workshop in Japan, Kanagawa, Japan, pp. 56–65 (1996)

52. Burns, K.: Style in poker. In: Proceedings of the 2006 IEEE Symposium on Computational Intelligence and Games (CIG 2006), Reno, NV, pp. 257–264. IEEE Press, Los Alamitos (2006)

53. Buro, M.: Probcut: An effective selective extension of the alpha-beta algorithm. ICCA Journal 18(2), 71–76 (1995)

54. Buro, M.: From simple features to sophisticated evaluation functions. In: van den Herik, H.J., Iida, H. (eds.) CG 1998. LNCS, vol. 1558, pp. 126–145. Springer, Heidelberg (1999)

55. Buro, M.: Toward opening book learning. ICCA Journal 22(2), 98–102 (1999)

56. Buro, M.: Improving heuristic mini-max search by supervised learning. Artificial Intelligence 134, 85–99 (2002)

57. Campbell, M., Hoane Jr., A.J., Hsu, F.-h.: Deep Blue. Artificial Intelligence 134, 57–83 (2002)

58. Caruana, R.: Multitask learning. Machine Learning 28, 41–75 (1997)

59. Charness, N., Reingold, E.M., Pomplun, M., Stampe, D.M.: The perceptual aspect of skilled performance in chess: Evidence from eye movements. Memory & Cognition 29(8), 1146–1152 (2001)

60. Chase, W.G., Simon, H.A.: The mind's eye in chess. In: Chase, W.G. (ed.) Visual information processing, pp. 215–281. Academic Press, New York (1973)

61. Chase, W.G., Simon, H.A.: Perception in chess. Cognitive Psychology 4, 55–81 (1973)

62. Chellapilla, K., Fogel, D.B.: Evolution, neural networks, games, and intelligence. Proceedings of the IEEE 87(9), 1471–1496 (1999)

63. Chellapilla, K., Fogel, D.B.: Evolving neural networks to play checkers without relying on expert knowledge. IEEE Transactions on Neural Networks 10(6), 1382–1391 (1999)

64. Chellapilla, K., Fogel, D.B.: Anaconda defeats Hoyle 6-0: A case study competing an evolved checkers program against commercially available software. In: Congress on Evolutionary Computation, La Jolla, CA, pp. 857–863 (2000)

65. Chellapilla, K., Fogel, D.B.: Evolving a neural network to play checkers without human expertise. In: Baba, N., Jain, L.C. (eds.) Computational Intelligence in Games, vol. 62, pp. 39–56. Springer, Berlin (2001)

66. Chellapilla, K., Fogel, D.B.: Evolving an expert checkers playing program without using human expertise. IEEE Transactions on Evolutionary Computation 5(4), 422–428 (2001)

67. Chen, X., Zhang, D., Zhang, X., Li, Z., Meng, X., He, S., Hu, X.: A functional MRI study of high-level cognition. II. The game of GO. Cognitive Brain Research 16, 32–37 (2003)

68. ChessBase - Chess Online Database (2005), http://www.chesslive.de/

69. Chikun, C. (ed.): Go: A Complate Introduction to the Gamenetworks for Optimization and Control. Kiseido Publishing Company (1997)

70. Chong, S.Y., Tan, M.K., White, J.D.: Observing the evolution of neural networks learning to play the game of Othello. IEEE Transactions on Evolutionary Computation 9(3), 240–251 (2005)

71. Clune, J.: Heuristic evaluation functions for General Game Playing. In: Proceedings of the Twenty-Second AAAI Conference on Artificial Intelligence (AAAI 2007), Vancouver, BC, Canada, pp. 1134–1139. AAAI Press, Menlo Park (2007)

72. Coulom, R.: Efficient selectivity and backup operators in Monte-Carlo tree search. In: van den Herik, H.J., Ciancarini, P., Donkers, H.H.L.M(J.) (eds.) CG 2006. LNCS, vol. 4630, pp. 72–83. Springer, Heidelberg (2007)

73. Cybenko, G.: Approximations by superpositions of sigmoidal functions. Mathematics of Control, Signals, and Systems 2(4), 303–314 (1989)

74. Dartnall, T. (ed.): Artificial Intelligence and Creativity: An Interdisciplinary Approach. Studies in Cognitive Systems, vol. 17. Springer, Heidelberg (1994)

75. Darwen, P., Yao, X.: Speciation as automatic categorical modularization. IEEE Transactions on Evolutionary Computation 1(2), 101–108 (1997)

76. Darwen, P.J.: Why co-evolution beats temporat diffrence learning at backgammon for a linear arcitecture, but not a non-linear architecture. In: Proceedings of the 2001 Congress on Evolutionary Computation (CEC 2001), pp. 1003–1010 (2001)

77. Davidson, A., Billings, D., Schaeffer, J., Szafron, D.: Improved opponent modeling in poker. In: International Conference on Artificial Intelligence (ICAI 2000), Las Vegas, NV, pp. 1467–1473 (2000)

78. de Groot, A.D.: Thought and Choice in Chess. Mouton Publishers, The Hague (1965); (Original work published in 1946 under the title Het denken van den schakar)

79. de Groot, A.D.: Thought and Choice in Chess, 2nd edn. Mouton Publishers, The Hague (1978)

80. de Groot, A.D., Gobet, F.: Perception and memory in chess, Van Gorcum, Assen, the Netherlands (2002)

81. Dendek, C., Mańdziuk, J.: Including metric space topology in neural networks training by ordering patterns. In: Kollias, S.D., Stafylopatis, A., Duch, W., Oja, E. (eds.) ICANN 2006. LNCS, vol. 4132, pp. 644–653. Springer, Heidelberg (2006)

82. Dendek, C., Mańdziuk, J.: A neural network classifier of chess moves. In: Proceedings of 8th Hybrid Intelligent Systems Conference (HIS 2008), Barcelona, Spain, pp. 338–343. IEEE Press, Los Alamitos (2008)

83. Dorigo, M., Birattari, M., Stutzle, T.: Ant Colony Optimization: Artificial ants as a Computational Intelligence technique. Technical Report TR/IRIDIA/2006-023, Universit Libre de Bruxelles (September 2006)

84. Duch, W.: Computational creativity. In: Proceedings of the World Congress on Computational Intelligence (WCCI 2006), Vancouver, Canada, pp. 1162–1169 (2006)

85. Duch, W.: Intuition, insight, imagination and creativity. IEEE Computational Intelligence Magazine 2(3), 40–52 (2007)

86. Dybicz, P.: International Bridge Master and trainer. Private communication (2008)

87. Eberhart, R.C., Shi, Y., Kennedy, J.: Swarm Intelligence. The Morgan Kaufmann Series in Artificial Intelligence. Morgan Kaufmann, San Francisco (2001)

88. Edwards, D.J., Hart, T.P.: The alpha-beta heuristic. Technical Report AI Memo - 30 (revised version), Massachusetts Institute of Technology (1963)

89. Elman, J.L.: Finding structure in time. Cognitive Science 14, 179–211 (1990)

90. Enzenberger, M.: The integration of a priori knowledge into a Go playing neural network. Technical report (1996)

91. Enzenberger, M.: Evaluation in Go by a neural network using soft segmentation. In: Advances in Computer Games: Many Games, Many Challenges: Proceedings of the International Conference on Advances in Computer Games (ACG-10), Graz, Austria, pp. 97–108 (2003)

92. Enzenberger, M., Müller, M.M.: A lock-free multithreaded monte-carlo tree search algorithm. In: Advances in Computer Games (ACG-12), Pamplona, Spain (2009)

93. Epstein, S.L.: The intelligent novice - learning to play better. In: Levy, D.N.L., Beal, D.F. (eds.) Heuristic Programming in Artificial Intelligence. The First Computer Olympiad. Ellis Horwood (1989)

94. Epstein, S.L.: Identifying the right reasons: Learning to filter decision makers. In: Greiner, R., Subramanian, D. (eds.) Proceedings of the AAAI 1994 Fall Symposium on Relevance, New Orleans, pp. 68–71. AAAI Press, Menlo Park (1994)

95. Epstein, S.L.: Toward an ideal trainer. Machine Learning 15(3), 251–277 (1994)

96. Epstein, S.L.: Game playing: The next moves. In: Proceedings of the Sixteenth National Conference on Artificial Intelligence, Orlando, FL, pp. 987–993 (1999)

97. Epstein, S.L.: Learning to play expertly: A tutorial on Hoyle. In: Fürnkrantz, J., Kubat, M. (eds.) Machines that learn to play games, pp. 153–178. Nova Science, Huntington (2001)

98. Epstein, S.L., Gelfand, J., Lesniak, J.: Pattern-based learning and spatially-oriented concept formation in a multi-agent, decision-making expert. Computational Intelligence 12(1), 199–221 (1996)

99. Epstein, S.L., Gelfand, J., Lock, E.: Learning game-specific spatially-oriented heuristics. Constraints 3(2-3), 239–253 (1998)

100. Fahlman, S.E.: An empirical study of learning speed in back-propagation networks. Technical report, Carnegie-Mellon University (1988)

101. Fawcett, T.E.: Feature discovery for problem solving systems. PhD thesis, University of Massachusetts at Amherst, Amherst, MA (1993)

102. Fawcett, T.E.: Knowledge-based feature discovery for evaluation functions. Computational Intelligence 11(4), 42–64 (1995)

103. Fawcett, T.E., Utgoff, P.E.: A hybrid method for feature generation. In: Machine Learning: Proceedings of the Eighth International Workshop, Evanston, IL, pp. 137–141. Morgan Kaufmann, San Francisco (1991)

104. Fawcett, T.E., Utgoff, P.E.: Automatic feature generation for problem solving systems. In: Sleeman, D., Edwards, P. (eds.) Proceedings of the 9th International Conference on Machine Learning, pp. 144–153. Morgan Kaufmann, San Francisco (1992)

105. Fietz, H.: Beyond the 3000 Elo barrier. a glance behind the scenes of the Rybka chess engine. Chess Magazine, 18–21 (May 2007)

106. Finkelstein, L., Markovitch, S.: Learning to play chess selectively by acquiring move patterns. International Computer Chess Association Journal 21, 100–119 (1998)

107. Finnsson, H., Björnsson, Y.: Simulation-based approach to General Game Playing. In: Proceedings of the Twenty-Third AAAI Conference on Artificial Intelligence (AAAI 2008), Chicago, IL, pp. 259–264. AAAI Press, Menlo Park (2008)

108. Fleuret, F.: Fast binary feature selection with conditional mutual information. Journal of Machine Learning Research 5, 1531–1555 (2004)

109. Fogel, D.B.: Blondie24: Playing at the Edge of Artificial Intelligence. Morgan Kaufmann, San Francisco (2001)

110. Fogel, D.B.: Evolutionary Computation. John Wiley & Sons, Chichester (2005)

111. Fogel, D.B., Chellapilla, K.: Verifying Anaconda's expert rating by competing against Chinook: experiments in co-evolving a neural checkers player. Neurocomputing 42, 69–86 (2002)

112. Fogel, D.B., Hays, T.J., Hahn, S.L., Quon, J.: A self-learning evolutionary chess program. Proceedings of the IEEE 92(12), 1947–1954 (2004)

113. Fogel, D.B., Hays, T.J., Hahn, S.L., Quon, J.: Further evaluation of a self-learning chess program. In: Kendall, G., Lucas, S.M. (eds.) Proceedings of the 2005 IEEE Symposium on Computational Intelligence and Games (CIG 2005), Essex, Great Britain, pp. 73–77. IEEE Press, Los Alamitos (2005)

114. Fogel, D.B., Hays, T.J., Hahn, S.L., Quon, J.: The Blondie25 chess program competes against Fritz 8.0 and a human chess master. In: Proceedings of the 2006 IEEE Symposium on Computational Intelligence and Games (CIG 2006), Reno, NV, pp. 230–235. IEEE Press, Los Alamitos (2006)

115. Franken, C.J.: PSO-based coevolutionary game learning. Master Thesis. Faculty of Engineering, Built-Environment and Information Technology, University of Pretoria, Pretoria, South Africa (2004)

116. Frayn, C.M., Justiniano, C., Lew, K.: ChessBrain II - a hierarchical infrastructure for distributed inhomogenous speed-critical computer. In: Proceedings of the 2006 IEEE Symposium on Computational Intelligence and Games (CIG 2006), Reno, NV, pp. 13–18. IEEE Press, Los Alamitos (2006)

117. French, R.M.: Catastrophic forgetting in connectionist networks: Causes, consequences and solutions. Trends in Cognitive Sciences, 128–135 (1994)

118. Fukushima, K., Miyake, S.: Neocognitron: a new algorithm for pattern recognition tolerant of deformations and shifts in position. Pattern Recognition 15(6), 455–469 (1982)

119. Fullmer, B., Miikkulainen, R.: Using marker-based genetic encoding of neural networks to evolve finite-state behaviour. In: Varela, F.J., Bourgine, P. (eds.) Towards a Practice of Autonomous Systems. Proceedings of the First European Conference on Artificial Life (ECAL 1991), pp. 255–262. MIT Press, Cambridge (1991)

120. Fürnkranz, J.: Machine learning in computer chess: the next generation. ICGA Journal 19(3), 147–161 (1996)

121. Fürnkranz, J.: Machine learning in games: A survey. In: Fürnkranz, J., Kubat, M. (eds.) Machines that Learn to Play Games, pp. 11–60. Nova Science Publishers, Huntington (2001)

122. Fürnkranz, J.: Recent advances in machine learning and game playing. ÖGAI Journal 26(2) (2007)

123. Gambäck, B., Rayner, M.: Contract Bridge as a micro-world for reasoning about communication agents. Technical Report SICS/R-90/9011, Swedish Institute of Computer Science (1990)

124. Gambäck, B., Rayner, M., Pell, B.: Pragmatic reasoning in Bridge. Technical Report 299, University of Cambridge, Computer Laboratory (April 1993)

125. Gelly, S., Wang, Y.: Exploration exploitation in Go: UCT for Monte-Carlo Go. In: Neural Information Processing Systems 2006 Workshop on On-line trading of exploration and exploitation (2006)

126. Gelly, S., Wang, Y., Munos, R., Teytaud, O.: Modification of UCT with patterns on Monte Carlo Go. Technical Report 6062, INRIA (2006)

127. Geman, S., Geman, D.: Stochastic relaxation, Gibbs distributions, and the Bayesian restoration of images. IEEE Transactions on Pattern Analysis and Machine Intelligence 6, 721–741 (1984)

128. Geman, S., Hwang, C.-R.: Diffusions for global optimization. SIAM Journal of Control and Optimization 24(4), 1031–1043 (1986)

129. Genesereth, M., Love, N.: General Game Playing: Overview of the AAAI Competition (2005), http://games.stanford.edu/aaai.pdf

130. Genesereth, M., Love, N., Pell, B.: General Game Playing: Overview of the AAAI Competition. AI Magazine 26(2), 62–72 (2005)

131. Gherrity, M.: A Game-Learning Machine. PhD thesis, University of California, San Diego, CA (1993)

132. Gillogly, J.: Performance analysis of the technology chess program. PhD thesis, Department of Computer Science, Carnagie-Mellon University, Pittsburgh, PA (1978)

133. Ginsberg, M.L.: GIB Library, http://www.cirl.uoregon.edu/ginsberg/gibresearch.html

134. Ginsberg, M.L.: GIB: Steps toward an expert-level bridge-playing program. In: International Joint Conference on Artificial Intelligence (IJCAI 1999), Stockholm, Sweden, pp. 584–589 (1999)

135. Ginsberg, M.L.: GIB: Imperfect information in a computationally challenging game. Journal of Artificial Intelligence Research 14, 303–358 (2001)

136. Givens, H.: PokerProbot, http://www.pokerprobot.com/

137. Gobet, F.: A computer model of chess memory. In: Proceedings of the 15th Annual Conference of the Cognitive Science Society, Boulder, CO, USA, pp. 463–468. Lawrence Erlbaum Associates, Mahwah (1993)

138. Gobet, F., Jansen, P.: Towards a chess program based on a model of human memory. In: van den Herik, H.J., Herschberg, I.S., Uiterwijk, J.W.H.M. (eds.) Advances in Computer Chess 7, pp. 35–60. University of Limburg (1994)

139. Gobet, F., Simon, H.A.: Five seconds or sixty? presentation time in expert's memory. Cognitive Psychology 24, 651–682 (2000)

140. Goldberg, D.E.: Genetic Algorithms in Search, Optimization and Machine Learning. Addison-Wesley, Reading (1989)

141. Gould, J., Levinson, R.: Experience-based adaptive search. In: Michalski, R., Tecuci, G. (eds.) Machine Learning: A Multi-Strategy Approach, pp. 579–604. Morgan Kaufmann, San Francisco (1994)

142. Graves, A., Fernández, S., Schmidhuber, J.: Multidimensional recurrent neural networks. In: de Sá, J.M., Alexandre, L.A., Duch, W., Mandic, D.P. (eds.) ICANN 2007. LNCS, vol. 4668, pp. 549–558. Springer, Heidelberg (2007)

143. Greer, K.: Computer chess move-ordering schemes using move influence. Artificial Intelligence 120, 235–250 (2000)
144. Greer, K.R.C., Ojha, P.C., Bell, D.A.: A pattern-oriented approach to move ordering: The chessmap heuristic. ICCA Journal 22(1), 13–21 (1999)
145. Hall, M.T., Fairbairn, J.: GoGoD Database and Encyclopaedia (2008), http://www.gogod.co.uk
146. Hassoun, M.H.: Fundamentals of Artificial Neural Networks. MIT Press, Cambridge (1995)
147. Haykin, S.: Neural Networks: A Comprehensive Foundation. Prentice-Hall, Englewood Cliffs (1998)
148. Heinz, E.A.: Adaptive null-move pruning. ICCA Journal 22(3), 123–132 (1999)
149. Hofstadter, D., Farg: Fluid Concepts and Creative Analogies: Computer Models of the Fundamental Mechanisms of Thought. Basic Books, New York (1995)
150. Holland, J.H.: Adaptation in Natural and Artificial Systems: An Introductory Analysis with Application to Biology, Control and Artificial Intelligence. University of Michigan Press, Ann Arbor (1975)
151. Hopfield, J.J.: Neurons with graded response have collective computational properties like those of two-state neurons. Proceedings of the National Academy of Science USA 81, 3088–3092 (1984)
152. Hopfield, J.J., Tank, D.W.: "Neural" computation of decisions in optimization problems. Biological Cybernetics 52, 141–152 (1985)
153. Hornik, K.: Approximation capabilities of multilayer feedforward networks. Neural Networks 4(2), 251–257 (1991)
154. Hornik, K., Stinchcombe, M., White, H.: Multilayer feedforward networks are universal approximators. Neural Networks 2(5), 359–366 (1989)
155. Hsu, F.-h.: Behind Deep Blue. Princeton University Press, Princeton (2002)
156. Hughes, E.: Piece difference: Simple to evolve? In: Proceedings of the 2003 Congress on Evolutionary Computation (CEC 2003), pp. 2470–2473 (2003)
157. Hughes, E.: Coevolving game strategies: How to win and how to lose. IEEE CIG 2005 Tutorial (2005), http://cswww.essex.ac.uk/cig/2005/
158. Hyatt, R.M.: Crafty (2006), ftp.cis.uab.edu/pub/hyatt
159. Hyatt, R.M., Nelson, H.L., Gower, A.E.: Cray Blitz. In: Marsland, T.A., Schaeffer, J. (eds.) Computers, Chess, and Cognition, pp. 111–130. Springer, New York (1990)
160. IBM Corporation. Deep Blue technology (2006), http://www.research.ibm.com/know/blue.html
161. Jollife, I.T.: Principal Component Analysis. Springer, New York (1986)
162. Jordan, M.I.: Serial order: a parallel distributed processing approach. Technical Report 8604, Institute for Cognitive Science, University of California, San Diego, La Jolla, CA (1986)
163. Kaelbling, L., Littman, M.L., Moore, A.W.: Reinforcement learning: A survey. Journal of Artificial Intelligence Research 4, 237–285 (1996)
164. Kaindl, H.: Tree search algorithms. In: Marsland, T.A., Schaeffer, J. (eds.) Computers, Chess, and Cognition, pp. 133–168. Springer, New York (1990)
165. Kalles, D., Kanellopoulos, P.: On verifying game designs and playing strategies using reinforcement learning. In: Proceedings of the 2001 ACM Symposium on Applied Computing (SAC 2001), Las Vegas, NV. ACM, New York (2001)
166. Katz-Brown, J., O'Laughlin, J.: Quackle - kwak!, http://www.quackle.org/

167. Kim, K.-J., Choi, H., Cho, S.-B.: Hybrid of evolution and reinforcement learning for Othello players. In: Proceedings of the 2007 IEEE Symposium on Computational Intelligence and Games (CIG 2007), Honolulu, Hawaii, pp. 203–209. IEEE Press, Los Alamitos (2007)

168. Kirkpatrick, S., Gelett Jr., C., Vecchi, M.: Optimization by simulated annealing. Science 220, 671–680 (1983)

169. Knuth, D.E., Moore, R.W.: An analysis of alpha-beta algorithm. Artificial Intelligence 6, 293–326 (1975)

170. Kocsis, L., Szepesvari, C.: Bandit based monte-carlo planning. In: Fürnkranz, J., Scheffer, T., Spiliopoulou, M. (eds.) ECML 2006. LNCS (LNAI), vol. 4212, pp. 282–293. Springer, Heidelberg (2006)

171. Kohonen, T.: Self-Organizing Maps, 3rd edn. Springer Series in Information Sciences, vol. 30. Springer, Heidelberg (2001)

172. Koller, D., Pfeffer, A.: Representations and solutions for game theoretic problems. Artificial Intelligence 94(1), 167–215 (1997)

173. Kopec, D., Bratko, I.: The Bratko-Kopec experiment: A comparison of human and computer performance in chess. In: Clarke, M.R.B. (ed.) Advances in Computer Chess 3, pp. 57–72. Pergamon Press, Oxford (1982)

174. Kotnik, C., Kalita, J.K.: The significance of Temporal-Difference learning in self-play training TD-Rummy versus EVO-Rummy. In: Fawcett, T., Mishra, N. (eds.) Machine Learning, Proceedings of the Twentieth International Conference (ICML 2003), Washington, DC, pp. 369–375. AAAI Press, Menlo Park (2003)

175. Kuhlmann, G., Dresner, K., Stone, P.: Automatic heuristic construction in a complete General Game Player. In: Proceedings of the Twenty-First AAAI Conference on Artificial Intelligence (AAAI 2006), Boston, MA, pp. 1457–1462. AAAI Press, Menlo Park (2006)

176. Kuijf, H.: Jack - computer bridge playing program, http://www.jackbridge.com

177. Kusiak, M., Walędzik, K., Mańdziuk, J.: Evolution of heuristics for give-away checkers. In: Duch, W., Kacprzyk, J., Oja, E., Zadrożny, S. (eds.) ICANN 2005. LNCS, vol. 3697, pp. 981–987. Springer, Heidelberg (2005)

178. Kusiak, M., Walędzik, K., Mańdziuk, J.: Evolutionary approach to the game of checkers. In: Beliczynski, B., Dzielinski, A., Iwanowski, M., Ribeiro, B. (eds.) ICANNGA 2007. LNCS, vol. 4431, pp. 432–440. Springer, Heidelberg (2007)

179. Lazar, S.: Analysis of transposition tables and replacement schemes. Dep. of Comp. Sci. and Electrical Engineering University of Maryland Baltimore County (1995)

180. LeCun, Y., Boser, B., Denker, J., Henderson, D., Howard, R., Hubbard, W., Jackel, L.: Backpropagation applied to handwritten character recognition. Neural Computation 5, 541–551 (1989)

181. LeCun, Y., Boser, B., Denker, J., Henderson, D., Howard, R., Hubbard, W., Jackel, L.: Handwritten digit recognition with a back-propagation network. In: Touretzky, D.S. (ed.) Advances in Neural Information Processing Systems (NIPS 2), Denver, Colorado, USA, pp. 396–404. Morgan Kaufmann, San Mateo (1990)

182. Lee, K.-F., Mahajan, S.: The delelopment of a world class othello program. Artificial Intelligence 43, 21–36 (1990)

183. Leouski, A.: Learning of position evaluation in the game of Othello. Master's Project. Department of Computer Science, University of Massachusetts, MA (1995)

184. Leouski, A.V., Utgoff, P.E.: What a neural network can learn about Othello. Technical Report 96-10, University of Massachusetts, Amherst, MA (1996)

185. Levinson, R.A.: A self-learning, pattern-oriented chess program. ICCA Journal 12(4), 207–215 (1989)

186. Levinson, R.A.: Experience-based creativity. In: Dartnall, T. (ed.) Artificial Intelligence and Creativity: An Interdisciplinary Approach. Studies in Cognitive Systems, vol. 17, pp. 161–179. Springer, Heidelberg (1994)

187. Levinson, R.A.: MORPH II: A universal agent: Progress report and proposal. Technical Report UCSC-CRL-94-22, Jack Baskin School of Engineering, Department of Computer Science, University of California, Santa Cruz (1994)

188. Levinson, R.A., Snyder, R.: Adaptive pattern-oriented chess. In: Birnbaum, L., Collins, G. (eds.) Proceedings of the 8th International Workshop on Machine Learning, pp. 85–89. Morgan Kaufmann, San Francisco (1991)

189. Levy, D.: The million pound bridge program. In: Levy, D., Beal, D. (eds.) Heuristic Programming in Artificial Intelligence - First Computer Olymiad, pp. 93–105. Ellis Horwood, Asilomar (1989)

190. Linhares, A.: An active symbols theory of chess intuition. Minds and Machines 15, 131–181 (2005)

191. Livingstone, D.: Perudish: a game framework and modified rule-set for Perudo. Computing and Information Systems Journal 9(3) (2005)

192. Livingstone, D.J.: Artificial Neural Networks. Methods and Applications. Methods in Molecular Biology, vol. 458. Springer, Heidelberg (2009)

193. Love, N., Genesereth, M., Hinrichs, T.: General game playing: Game description language specification. Technical Report LG-2006-01, Stanford University, Stanford, CA (2006),
http://logic.stanford.edu/reports/LG-2006-01.pdf

194. Lubberts, A., Miikkulainen, R.: Co-evolving a Go-playing neural network. In: Coevolution: Turning Adaptive Algorithms upon Themselves, Birds-of-a-Feather Workshop, Genetic and Evolutionary Computation Conference (Gecco 2001), San Francisco, USA, pp. 14–19 (2001)

195. Lucas, S.M.: Computational intelligence and games: Challenges and opportunities. International Journal of Automation and Computing 5, 45–57 (2008)

196. Lucas, S.M.: Learning to play othello with n-tuple systems. Australian Journal of Intelligent Information Processing 4, 1–20 (2008)

197. Lucas, S.M., Kendall, G.: Evolutionary computation and games. IEEE Computational Intelligence Magazine, 10–18 (February 2006)

198. Lucas, S.M., Runarsson, T.P.: Othello Competition (2006),
http://algoval.essex.ac.uk:8080/othello/html/Othello.html

199. Lucas, S.M., Runarsson, T.P.: Temporal Difference Learning versus Coevolution for acquiring Othello position evaluation. In: Proceedings of the 2006 IEEE Symposium on Computational Intelligence and Games, Reno, NV, pp. 52–59. IEEE Press, Los Alamitos (2006)

200. Macleod, A.: Perudo as a development platform for Artificial Intelligence. In: 13th Game-On International Conference (CGAIDE 2004), Reading, UK, pp. 268–272 (2004)

201. Macleod, A.: Perudo game (2006), http://www.playperudo.com/

202. Malmgren, H., Borga, M., Niklasson, L. (eds.): Artificial Neural Networks in Medicine and Biology. Proceedings of the ANNIMAB-1 Conference. Perspectives in Neural Computing. Springer, Göteborg (2000)

203. Mańdziuk, J.: Incremental learning approach for board game playing agents. In: Proceedings of the 2000 International Conference on Artificial Intelligence (ICAI 2000), Las Vegas, vol. 2, pp. 705–711 (2000)

204. Mańdziuk, J.: Incremental training in game playing domain. In: Proceedings of the International ICSC Congress on Intelligent Systems & Applications (ISA 2000), Wollongong, Australia, vol. 2, pp. 18–23 (2000)

205. Mańdziuk, J. (ed.): Neural networks for Optimization and Control. Special Issue of Control and Cybernetics (2002)

206. Mańdziuk, J.: Computational Intelligence in Mind Games. In: Duch, W., Mańdziuk, J. (eds.) Challenges for Computational Intelligence. Studies in Computational Intelligence, vol. 63, pp. 407–442. Springer, Heidelberg (2007)

207. Mańdziuk, J.: Some thoughts on using Computational Intelligence methods in classical mind board games. In: Procedings of the 2008 International Joint Conference on Neural Networks (IJCNN 2008), Hong Kong, China, pp. 4001–4007 (2008)

208. Mańdziuk, J., Kusiak, M., Walędzik, K.: Evolutionary-based heuristic generators for checkers and give-away checkers. Expert Systems 24(4), 189–211 (2007)

209. Mańdziuk, J., Mossakowski, K.: Looking inside neural networks trained to solve double-dummy bridge problems. In: 5th Game-On International Conference on Computer Games: Artificial Intelligence, Design and Education (CGAIDE 2004), Reading, UK, pp. 182–186 (2004)

210. Mańdziuk, J., Mossakowski, K.: Example-based estimation of hand's strength in the game of bridge with or without using explicit human knowledge. In: Proceedings of the IEEE Symposium on Computational Intelligence in Data Mining (CIDM 2007), Honolulu, Hawaii, pp. 413–420. IEEE Press, Los Alamitos (2007)

211. Mańdziuk, J., Mossakowski, K.: Neural networks compete with expert human bridge players in solving the Double Dummy Bridge Problem. In: Proceedings of the 2009 IEEE Symposium on Computational Intelligence and Games (CIG 2009), Milan, Italy, pp. 117–124. IEEE Press, Los Alamitos (2009)

212. Mańdziuk, J., Osman, D.: Alpha-beta search enhancements with a real-value game state evaluation function. ICGA Journal 27(1), 38–43 (2004)

213. Mańdziuk, J., Osman, D.: Temporal difference approach to playing give-away checkers. In: Rutkowski, L., Siekmann, J.H., Tadeusiewicz, R., Zadeh, L.A. (eds.) ICAISC 2004. LNCS (LNAI), vol. 3070, pp. 909–914. Springer, Heidelberg (2004)

214. Mańdziuk, J., Shastri, L.: Incremental Class Learning approach and its application to handwritten digit recognition. Information Sciences 141(3-4), 193–217 (2002)

215. Manning, E.: Temporal Difference learning of an Othello evaluation function for a small neural network with shared weights. In: Proceedings of the 2007 IEEE Symposium on Computational Intelligence and Games (CIG 2007), Honolulu, Hawaii, pp. 216–223. IEEE Press, Los Alamitos (2007)

216. Marshall, J.: Metacat: A Self-watching Cognitive Architecture for Analogy-making and High Level Perception. PhD thesis, Indiana University, Bloomington (1999)
217. Marsland, T.A.: Relative efficiency of alpha-beta implementations. In: Proceedings of the International Joint Conference on Artificial Intelligence, Karlsruhe, pp. 763–766 (1983)
218. Marsland, T.A.: Evaluation-function factors. ICCA Journal 8(2), 47–57 (1985)
219. Marsland, T.A., Campbell, M.: Parallel search of strongly ordered game trees. ACM Computing Surveys 14(4), 533–551 (1982)
220. Mayer, H.: Board representations for neural Go players learning by Temporal Difference. In: Proceedings of the 2007 IEEE Symposium on Computational Intelligence and Games (CIG 2007), pp. 183–188. IEEE Press, Los Alamitos (2007)
221. McAllester, D.: Conspiracy numbers for min-max search. Artificial Intelligence 35, 287–310 (1988)
222. McCarthy, J.: Homepage of John McCarthy (1998), http://www-formal.stanford.edu/jmc/reti.html
223. McCloskey, M., Cohen, N.J.: Catastrophic interference in connectionist networks: The sequential learning problem. In: Bower, G.H. (ed.) The Psychology of Learning and Motivation, vol. 24, pp. 109–164. Academic Press, San Diego (1989)
224. McCulloch, W.S., Pitts, W.: A logical calculus of the ideas immanent in nervous activity. Mathematical Biophysics 5, 115–133 (1943)
225. McNelis, P.D.: Neural Networks in Finance: Gaining Predictive Edge in the Market. Elsevier Academic Press, Amsterdam (2004)
226. Mitchell, M.: Analogy-making as Perception. MIT Press, Cambridge (1993)
227. Mitchell, M., Hofstadter, D.R.: The emergence of understanding in a computer model of concepts and analogy making. Physica D 42, 322–334 (1990)
228. Mitchell, T.M., Thrun, S.: Explanation based learning: A comparison of symbolic and neural network approaches. In: Utgoff, P.E. (ed.) Proceedings of the 10th International Conference on Machine Learning, San Mateo, CA, pp. 197–204. Morgan Kaufmann, San Francisco (1993)
229. Miwa, M., Yokoyama, D., Chikayama, T.: Automatic generation of evaluation features for computer game players. In: Proceedings of the 2007 IEEE Symposium on Computational Intelligence and Games (CIG 2007), Honolulu, Hawaii, pp. 268–275. IEEE Press, Los Alamitos (2007)
230. Moriarty, D.E., Miikkulainen, R.: Evolving neural networks to focus minimax search. In: Proceedings of the Twelfth National Conference on Artificial Intelligence (AAAI 1994), Seattle, WA, pp. 1371–1377. MIT Press, Cambridge (1994)
231. Moriarty, D.E., Miikkulainen, R.: Discovering complex Othello strategies through evolutionary neural systems. Connection Science 7(3), 195–209 (1995)
232. Moriarty, D.E., Miikkulainen, R.: Forming neural networks through efficient and adaptive coevolution. Evolutionary Computation 5, 373–399 (1997)
233. Mossakowski, K., Mańdziuk, J.: Artificial neural networks for solving double dummy bridge problems. In: Rutkowski, L., Siekmann, J.H., Tadeusiewicz, R., Zadeh, L.A. (eds.) ICAISC 2004. LNCS (LNAI), vol. 3070, pp. 915–921. Springer, Heidelberg (2004)

234. Mossakowski, K., Mańdziuk, J.: Neural networks and the estimation of hands' strength in contract bridge. In: Rutkowski, L., Tadeusiewicz, R., Zadeh, L.A., Żurada, J.M. (eds.) ICAISC 2006. LNCS (LNAI), vol. 4029, pp. 1189–1198. Springer, Heidelberg (2006)

235. Mossakowski, K., Mańdziuk, J.: Learning without human expertise. a case study of the double dummy bridge problem. IEEE Transactions on Neural Networks 20(2), 278–299 (2009)

236. Müller, M.: Computer Go as a sum of local games: An application of combinatorial game theory. PhD thesis, ETH Zürich, Switzerland (1995)

237. Müller, M.: Computer Go. Artificial Intelligence 134, 145–179 (2002)

238. Newell, A., Shaw, J.C., Simon, H.A.: Chess-playing programs and the problem of complexity. IBM Journal of Research and Development 2(4), 320–335 (1958)

239. Nikravesh, M., Kacprzyk, J., Zadeh, L.A. (eds.): *Forging New Frontiers: Fuzzy Pioneers I*. Studies in Fuzziness and Soft Computing, vol. 217. Springer, Heidelberg (2007)

240. Osaki, Y., Shibahara, K., Tajima, Y., Kotani, Y.: An Othello evaluation function based on Temporal Difference learning using probability of winning. In: Proceedings of the 2008 IEEE Symposium on Computational Intelligence and Games (CIG 2008), Perth, Australia, pp. 205–211. IEEE Press, Los Alamitos (2008)

241. Osman, D.: Effectiveness of TD-learning methods in two-player board games. PhD thesis, Systems Research Institute, Polish Academy of Science, Warsaw, Poland (2007) (in Polish)

242. Osman, D., Mańdziuk, J.: Comparison of tdleaf(λ) and td(λ) learning in game playing domain. In: Pal, N.R., Kasabov, N., Mudi, R.K., Pal, S., Parui, S.K. (eds.) ICONIP 2004. LNCS, vol. 3316, pp. 549–554. Springer, Heidelberg (2004)

243. Osman, D., Mańdziuk, J.: TD-GAC: Machine Learning experiment with give-away checkers. In: Dramiński, M., Grzegorzewski, P., Trojanowski, K., Zadrożny, S. (eds.) Issues in Intelligent Systems. Models and Techniques, Exit, pp. 131–145 (2005)

244. Papacostantis, E., Engelbrecht, A.P., Franken, N.: Coevolving probabilistic game playing agents using Particle Swarm Optimization algorithms. In: Proceedings of the 2005 IEEE Symposium on Computational Intelligence in Games (CIG 2005), Colchester, UK, pp. 195–202 (2005)

245. Pawlak, Z.: Rough Sets. International Journal of Computer and Information Sciences 11, 341–356 (1982)

246. Pawlak, Z.: Rough Sets. Theoretical Aspects of Reasoning about Data. Kluwer Academic Publishers, Dordrecht (1991)

247. Pearl, J.: SCOUT: A Simple Game-Searching Algorithm with Proven Optimal Properties. In: Proceedings of the 1st Annual National Conference on Artificial Intelligence (AAAI 1980), Stanford, CA, pp. 143–145 (1980)

248. Pell, B.: Metagame: a new challenge for games and learning. In: van den Herik, H.J., Allis, L.V. (eds.) Heuristic Programming in Artificial Intelligence. The Third Computer Olympiad, Ellis Horwood (1992)

249. Pell, B.: Metagame in symmetric chess-like games. In: van den Herik, H.J., Allis, L.V. (eds.) Heuristic Programming in Artificial Intelligence. The Third Computer Olympiad, Ellis Horwood (1992)

250. Pell, B.: Strategy Generation and Evaluation for Meta-Game Playing. PhD thesis, Computer Laboratory, University of Cambridge, UK (1993)

251. Pell, B.: A strategic metagame player for general chess-like games. In: 11th National Conference on Artificial Intelligence (AAAI 1994), pp. 1378–1385 (1994)

252. Pell, B.: A strategic metagame player for general chess-like games. Computational Intelligence 12, 177–198 (1996)

253. Plaat, A., Schaeffer, J., Pijls, W., de Bruin, A.: Best-first fixed-depth game-tree search in practice. In: Proceedings of the International Joint Conference on Artificial Intelligence (IJCAI 1995), Montreal, Quebec, Canada, vol. 1, pp. 273–279 (1995)

254. Plaat, A., Schaeffer, J., Pijls, W., de Bruin, A.: Exploiting graph properties of game trees. In: 13th National Conference on Artificial Intelligence (AAAI 1996), Menlo Park, CA, vol. 1, pp. 234–239 (1996)

255. Plaat, A., Schaeffer, J., Pijls, W., de Bruin, A.: A minimax algorithm better than SSS*. Artificial Intelligence 87(1-2), 255–293 (1996)

256. Poe, E.A.: Maelzel's chess player. Southern Literary Messenger (April 1936)

257. Pollack, J.B., Blair, A.D.: Why did TD-Gammon work? In: Mozer, M.C., Jordan, M.I., Petsche, T. (eds.) Advances in Neural Information Processing Systems (NIPS 9), Denver, CO, USA, pp. 10–16. MIT Press, Cambridge (1997)

258. Pollack, J.B., Blair, A.D.: Co-evolution in the successful learning of backgammon strategy. Machine Learning 32(3), 225–240 (1998)

259. Pollack, J.B., Blair, A.D., Land, M.: Coevolution of a backgammon player. In: Langton, C.G., Shimokara, K. (eds.) Proceedings of the Fifth Artificial Life Conference, pp. 92–98. MIT Press, Cambridge (1997)

260. Quenault, M., Cezaneve, T.: Extended general gaming model. In: van den Herik, H.J., Uiterwijk, J., Winands, M., Schadd, M. (eds.) Proceedings of the Computer Games Workshop (CGW 2007), Amsterdam, The Netherlands, pp. 195–204. MICC-IKAT (2007)

261. Rajlich, V.: Chess program Rybka, http://www.rybkachess.com/

262. Reinefeld, A.: An improvement to the scout tree-search algorithm. ICCA Journal 6(4), 4–14 (1983)

263. Reingold, E.M., Charness, N.: Perception in chess: Evidence from eye movements. In: Underwood, G. (ed.) Cognitive processes in eye guidance, pp. 325–354. Oxford University Press, Oxford (2005)

264. Reisinger, J., Bahçeci, E., Karpov, I., Miikkulainen, R.: Coevolving strategies for general game playing. In: Proceedings of the 2007 IEEE Symposium on Computational Intelligence and Games (CIG 2007), Honolulu, Hawaii, pp. 320–327. IEEE Press, Los Alamitos (2007)

265. Reitman, J.S.: Skilled perception in Go: Deducing memory structures from inter-response times. Cognitive Psychology 8(3), 336–356 (1976)

266. Richards, M., Amir, E.: Opponent modeling in scrabble. In: Veloso, M.M. (ed.) Proceedings of the 20th International Joint Conference on Artificial Intelligence (IJCAI 2007), Hyderabad, India, pp. 1482–1487 (2007)

267. Richards, N., Moriarty, D.E., Miikkulainen, R.: Evolving neural networks to play Go. Applied Intelligence 8, 85–96 (1998)

268. Riedmiller, M., Braun, H.: RPROP - a fast adaptive learning algorithm. Technical report, Universitat Karlsruhe (1992)

269. Riedmiller, M., Braun, H.: A direct adaptive method for faster backprop-agation: the RPROP algorithm. In: Proceedings of the IEEE International Conference on Neural Networks (ICNN 1993), pp. 586–591. IEEE Press, Los Alamitos (1993)

270. Rosenbloom, P.S.: A world-championship-level Othello program. Artificial Intelligence 19, 279–320 (1982)

271. Rosin, C.: Coevolutionary search among adversaries. PhD thesis, University of California, San Diego, CA (1997)

272. Rosin, C., Belew, R.: New methods for competitive coevolution. Evolutionary Computation 5(1), 1–29 (1997)

273. Rumelhart, D.E., Hinton, G.E., Williams, R.J.: Learning internal representa-tions by error backpropagation. In: Rumelhart, D.E., McClelland, J.L. (eds.) Parallel Distributed Processing, ch. 8, vol. 1, pp. 318–362. MIT Press, Cam-bridge (1986)

274. Rumelhart, D.E., Hinton, G.E., Williams, R.J.: Learning representations by back–propagating errors. Nature 323, 533–536 (1986)

275. Runarsson, T.P., Lucas, S.M.: Coevolution versus self-play temporal difference learning for acquiring position evaluation on small-board Go. IEEE Transac-tions on Evolutionary Computation 9(6), 628–640 (2005)

276. Russel, S., Norvig, P.: Artificial Intelligence: A Modern Approach, 2nd edn. Prentice Hall, Upper Sadle River (2003)

277. Samuel, A.L.: Some studies in machine learning using the game of checkers. IBM Journal of Research and Development 3(3), 210–229 (1959)

278. Samuel, A.L.: Some studies in machine learning using the game of checkers. In: Feigenbaum, E.A., Feildman, J. (eds.) Computers and Thought, pp. 71–105. McGraw-Hill, New York (1963)

279. Samuel, A.L.: Some studies in machine learning using the game of checkers II - recent progress. IBM Journal of Research and Development 11(6), 601–617 (1967)

280. Schaeffer, J.: Poki-X, http://www.cs.ualberta.ca/~games/poker/

281. Schaeffer, J.: The history heuristic. International Computer Chess Association Journal 6, 16–19 (1983)

282. Schaeffer, J.: The history heuristic and alpha-beta search enhancements in practice. IEEE PAMI 11(11), 1203–1212 (1989)

283. Schaeffer, J.: Conspiracy numbers. Artificial Intelligence 43, 67–84 (1990)

284. Schaeffer, J.: One Jump Ahead: Challenging Human Supremacy in Checkers. Springer, New York (1997)

285. Schaeffer, J.: Private communication (2005)

286. Schaeffer, J., Burch, N., Björnsson, Y., Kishimoto, A., Müller, M., Lake, R., Lu, P., Sutphen, S.: Checkers is solved. Science 317, 1518–1522 (2007)

287. Schaeffer, J., Culberson, J.C., Treloar, N., Knight, B., Lu, P., Szafron, D.: A world championship caliber checkers program. Artificial Intelligence 53(2-3), 273–289 (1992)

288. Schaeffer, J., Hlynka, M., Jussila, V.: Temporal difference learning applied to a high-performance game-playing program. In: International Joint Conference on Artificial Intelligence (IJCAI 2001), pp. 529–534 (2001)

289. Schaeffer, J., Lake, R., Lu, P., Bryant, M.: Chinook: The world man-machine checkers champion. AI Magazine 17(1), 21–29 (1996)

290. Schaul, T., Schmidhuber, J.: A scalable neural network architecture for board games. In: Proceedings of the 2008 IEEE Symposium on Computational Intelligence and Games (CIG 2008), Perth, Australia, pp. 357–364. IEEE Press, Los Alamitos (2008)

291. Schaul, T., Schmidhuber, J.: Scalable neural networks for board games. In: Alippi, C., Polycarpou, M., Panayiotou, C., Ellinas, G. (eds.) Artificial Neural Networks – ICANN 2009. LNCS, vol. 5768, pp. 1005–1014. Springer, Heidelberg (2009)

292. Schiffel, S., Thielscher, M.: Fluxplayer: A successful General Game Player. In: Proceedings of the Twenty-Second AAAI Conference on Artificial Intelligence (AAAI 2007), Vancouver, BC, Canada, 2007, pp. 1191–1196. AAAI Press, Menlo Park (2007)

293. Schraudolph, N.N., Dayan, P., Sejnowski, T.J.: Temporal difference learning of position evaluation in the game of Go. In: Cowan, J.D., Tesauro, G., Alspector, J. (eds.) Advances in Neural Information Processing, vol. 6, pp. 817–824. Morgan Kaufmann, San Francisco (1994)

294. Schraudolph, N.N., Dayan, P., Sejnowski, T.J.: Learning to evaluate Go positions via Temporal Difference learning. Technical Report 05–00, IDSIA (2000)

295. Schraudolph, N.N., Dayan, P., Sejnowski, T.J.: Learning to evaluate Go positions via Temporal Difference methods. In: Baba, N., Jain, L.C. (eds.) Computational Intelligence in Games, vol. 62, pp. 77–98. Springer, Berlin (2001)

296. Shannon, C.E.: Programming a computer for playing chess. Philosophical Magazine 41(7th series)(314), 256–275 (1950)

297. Sharma, S., Kobti, Z., Goodwin, S.: Learning and knowledge generation in general games. In: Proceedings of the 2008 IEEE Symposium on Computational Intelligence and Games (CIG 2008), Perth, Australia, 2008, pp. 329–335. IEEE Press, Los Alamitos (2008)

298. Sheppard, B.: World-championship-caliber scrabble. Artificial Intelligence 134, 241–275 (2002)

299. Simon, H.: Making management decisions: The role of intuition and emotion. In: Agor, W. (ed.) Intuition in Organizations, pp. 23–39. Sage Pubs., London (1987)

300. Simon, H.A.: Explaining the ineffable: AI on the topics of intuition, insights and inspiration. In: Proceedings of the 14th International Joint Conference on Artificial Intelligence (IJCAI 1995), vol. 1, pp. 939–948 (1995)

301. Simon, H.A., Chase, W.G.: Skill in chess. American Scientist 61(4), 394–403 (1973)

302. Singh, S.P., Sutton, R.S.: Reinforcement learning with replacing eligibility traces. Machine Learning 22(1-3), 123–158 (1996)

303. Skiena, S.S.: An overview of machine learning in computer chess. ICCA Journal 9(1), 20–28 (1986)

304. Sklansky, D.: Hold'Em Poker. Two Plus Two Publishing, Nevada (1996)

305. Sklansky, D., Malmuth, M.: Hold'Em Poker for Advanced Players, 21st Century edn. Two Plus Two Publishing, Nevada (2001)

306. Stanley, K.O.: Compositional pattern producing networks: A novel abstraction of development. Genetic Programming and Evolvable Machines 8(2), 131–162 (2007)

307. Stanley, K.O., Miikkulainen, R.: Evolving neural networks through augmenting topologies. Evolutionary Computation 10(2), 99–127 (2002)

308. Stern, D., Herbrich, R., Graepel, T.: Bayesian pattern ranking for move prediction in the game of Go. In: Proceedings of the 23rd International Conference on Machine Learning, Pittsburgh, PA, pp. 873–880 (2006)

309. Sternberg, R.J., Lubart, T.I.: The concepts of creativity: Prospects and paradigms. In: Sternberg, R.J. (ed.) Handbook of Creativity, pp. 3–15. Cambridge University Press, Cambridge (1999)

310. Stilman, B.: Liguistic Geometry. From search to construction. Kluwer Academic Publishers, Dordrecht (2000)

311. Stockman, G.: A minimax algorithm better than alfa-beta? Artificial Intelligence 12(2), 179–196 (1979)

312. Sutskever, I., Nair, V.: Mimicking Go experts with convolutional neural networks. In: Kůrková, V., Neruda, R., Koutník, J. (eds.) ICANN 2008, Part II. LNCS, vol. 5164, pp. 101–110. Springer, Heidelberg (2008)

313. Sutton, R.: Learning to predict by the methods of temporal differences. Machine Learning 3, 9–44 (1988)

314. Sutton, R.S., Barto, A.G.: Reinforcement Learning: An Introduction. MIT Press, Cambridge (1998)

315. Tesauro, G.: Connectionist learning of expert preferences by comparison training. In: Touretzky, D.S. (ed.) Advances in Neural Information Processing Systems (NIPS 1), Denver, Colorado, USA. Morgan Kaufmann, San Francisco (1989)

316. Tesauro, G.: Neurogammon wins computer olympiad. Neural Computation 1, 321–323 (1989)

317. Tesauro, G.: Practical issues in Temporal Difference Learning. Machine Learning 8, 257–277 (1992)

318. Tesauro, G.: TD-Gammon, a self-teaching backgammon program, achieves master-level play. Neural Computation 6(2), 215–219 (1994)

319. Tesauro, G.: Temporal Difference Learning and TD-Gammon. Communications of the ACM 38(3), 58–68 (1995)

320. Tesauro, G.: Comments on co-evolution in the successful learning of backgammon strategy. Machine Learning 32(3), 241–243 (1998)

321. Tesauro, G.: Programming backgammon using self-teaching neural nets. Artificial Intelligence 134, 181–199 (2002)

322. Tesauro, G., Galperin, G.R.: On-line policy improvement using Monte-Carlo search. In: Mozer, M.C., Jordan, M.I., Petsche, T. (eds.) Advances in Neural Information Processing Systems (NIPS 9), pp. 1068–1074 (1997)

323. Thrun, S.: Learning to play the game of chess. In: Tesauro, G., Touretzky, D., Leen, T. (eds.) Advances in Neural Information Processing Systems, vol. 7, pp. 1069–1076. The MIT Press, Cambridge (1995)

324. Thrun, S.: Explanation-Based Neural Network Learning: A Lifelong Learning Approach. Kluwer Academic Publishers, Boston (1996)

325. Thrun, S., Mitchell, T.M.: Learning one more thing. Technical report, Carnegie Mellon University, Pittsburg, CMU-CS-94-184 (1994)

326. Turing, A.M.: Computing machinery and intelligence. Mind 59(236), 433–460 (1950)

327. Turing, A.M.: Digital computers applied to games. In: Bowden, B.V. (ed.) Faster than thought: a symposium on digital computing machines, ch. 25, Pitman, London, UK (1953)

328. United States Chess Federation. Chess rules, http://www.uschess.org/

329. Utgoff, P.E.: Feature construction for game playing. In: Fürnkranz, J., Kubat, M. (eds.) Machines that Learn to Play Games, pp. 131–152. Nova Science Publishers, Huntington (2001)

330. van der Werf, E.C.D., Uiterwijk, J.W.H.M., Postma, E., van den Herik, H.J.: Local move prediction in Go. In: Proceedings of the Third International Conference on Computers and Games (CG 2002), Edmontion, Canada, pp. 393–412 (2002)

331. van der Werf, E.C.D., van den Herik, H.J., Uiterwijk, J.W.H.M.: Solving Go on small boards. International Computer Games Association Journal 26(2), 92–107 (2003)

332. von Neumann, J.: Zur theorie der gesellschaftsspiel. Math. Annalen 100, 295–320 (1928)

333. von Neumann, J., Morgenstern, O.: Theory of Games and Economic Behavior. Princeton Univ. Press, Princeton (1944)

334. Walczak, S.: Predicting actions from induction on past performance. In: Proceedings of the 8th International Workshop on Machine Learning, pp. 275–279. Morgan Kaufmann, San Francisco (1991)

335. Walczak, S.: Improving opening book performance through modeling of chess opponents. In: Proceedings of the 1996 ACM 24th Annual Conference on Computer Science, pp. 53–57. ACM, New York (1996)

336. Walczak, S.: Knowledge-based search in competitive domains. IEEE Transactions on Knowledge and Data Engineering 15(3), 734–743 (2003)

337. Walczak, S., Dankel, D.D.: Acquiring tactical and strategic knowledge with a generalized method for chunking of game pieces. International Journal of Intelligent Systems 8(2), 249–270 (1993)

338. Walker, S., Lister, R., Downs, T.: On self-learning patterns in the othello board game by the method of temporal differences. In: Proceedings of the 6th Australian Joint Conference on Artificial Intelligence, Melbourne, pp. 328–333. World Scientific, Singapore (1993)

339. Wang, D., Yuwono, B.: Incremental learning of complex temporal patterns. IEEE Transactions on Neural Networks 7(6), 1465–1481 (1996)

340. Wang, Y., Gelly, S.: Modifications of UCT and sequence-like simulations for Monte-Carlo Go. In: Proceedings of the 2007 IEEE Symposium on Computational Intelligence and Games (CIG 2007), Honolulu, Hawaii, pp. 175–182. IEEE Press, Los Alamitos (2007)

341. Watkins, C.J.C.H.: Learning from delayed rewards. PhD thesis, King's College, Cambridge, UK (1989)

342. Watkins, C.J.C.H., Dayan, P.: Q-learning. Machine Learning 8(3), 279–292 (1992)

343. WBridge5 - computer bridge playing program (in French), http://www.wbridge5.com/

344. Werbos, P.J.: Beyond regression: New Tools for Prediction and Analysis in the Behavioral Sciences. PhD thesis, Harvard University, Cambridge, MA (1974)

345. Wilcoxon, F.: Individual comparisons by ranking methods. Biometrics 1, 80–83 (1945)

346. Wilkins, D.: Using patterns and plans in chess. Artificial Intelligence 14, 165–203 (1980)

347. Winkler, F.-G., Fürnkranz, J.: On effort in AI research: A description along two dimensions. In: Morris, R. (ed.) Deep Blue versus Kasparov: The Significance for Artificial Intelligence: Papers from the 1997 AAAI Workshop, pp. 56–62. AAAI Press, Providence (1997)

348. Winkler, F.-G., Fürnkranz, J.: A hypothesis on the divergence of AI research. International Computer Chess Association Journal 21, 3–13 (1998)

349. Wolpert, D.H., Macready, W.G.: No Free Lunch theorems for optimization. IEEE Transactions on Evolutionary Computation 1(1), 67–82 (1997)

350. Yager, R.R., Filev, D.P.: Essentials of Fuzzy Modeling and Control. John Wiley & Sons, Inc., New York (1994)

351. Yao, X., Lu, Y., Darwen, P.: How to make best use of evolutionary learning. In: Stocker, R., Jelinck, H., Burnota, B., Maier, T.B. (eds.) Complex Systems - From Local Interactions to Global Phenomena, pp. 229–242. IOS Press, Amsterdam (1996)

352. Zadeh, L.A.: Fuzzy Sets. Information and Control 8(3), 338–353 (1965)

353. Zobrist, A.: Feature extractions and representation for pattern recognition and the game of Go. PhD thesis, University of Wisconsin (1970)

354. Zobrist, A.: A new hashing method with applications for game playing. ICCA Journal 13(2), 69–73 (1990)